PLANT SCIENCE RESEARCH AND PRACTICES

ASPERGILLUS NIGER

PATHOGENICITY, CULTIVATION AND USES

PLANT SCIENCE RESEARCH AND PRACTICES

Additional books and e-books in this series can be found on Nova's website under the Series tab.

PLANT SCIENCE RESEARCH AND PRACTICES

ASPERGILLUS NIGER

PATHOGENICITY, CULTIVATION AND USES

EVA BAUGHAN
EDITOR

nova
science publishers
New York

Library of Congress Cataloging-in-Publication Data

ISBN: 978-1-53618-080-0
Library of Congress Control Number: 2020939198

Published by Nova Science Publishers, Inc. † New York

CONTENTS

PREFACE

Aspergillus niger: Pathogenicity, Cultivation and Uses summarizes the results of numerous research experiments on the biomineralization of rocks and minerals carried out with the participation of the fungus *Aspergillus niger*. This compilation reviews recent advances in biocompounds and products with the most relevant biological and biocatalytic activities produced by *Aspergillus niger*. The challenges and future perspectives of biotechnological process extensions using *Aspergillus niger* will also be discussed. The closing study focuses on how *Aspergillus niger* may be utilized as a bioadsorbent for the detoxification and decolorizing of wastewater. In industrial and pharmaceutical applications, it can similarly be utilized in the creation of some costly and helpful biopolymers, such as chitin, chitosan, and nourishment enhancements like chitin-glucan and glucosamine.

Chapter 1 - The present review article summarizes the results of numerous research experiments on biomineralization on rocks and minerals surface carried out with the participation of fungus *Aspergillus niger*. The variations of fungus cultivation on liquid and solid media as well as in a moist chamber were reviewed. Metabolism of different strains of *A. niger* isolated from various habitats was discussed. The main attention was paid to the production of organic acids, especially oxalic acid, which plays a leading role in the interaction between fungus and rock.

The influence of cultivation conditions on the metabolic activity of *A. niger* and of *A. niger* + *Bacillus subtilis* association was analyzed. It was revealed that among the components of culture medium that catalyse or inhibit the development and the metabolism of *A. niger*, we can, first of all, highlight the concentration of sugar, the chemical form of nitrogen, as well as the presence of calcium carbonate and heavy metals in the medium. Metabolism of *A. niger* and of *A. niger* + *B. subtilis* association in the same conditions was vastly different which significantly affects bio-mineralization. The patterns of following oxalates formation on different mineral substrates by *A. niger* participation were revealed: calcium oxalates (weddellite and whewellite), strontium oxalates, copper oxalates (moolooite, fiemmeite, wheatleyite), iron oxalate (humboldtine), magnesium oxalate (glushinskite), manganese oxalates (lindbergite and falottaite) and finally lead and zinc oxalates which don't have mineral analogues yet met in nature – on cerussite and smitsonite. It was shown that morphogenetic patterns of oxalate formation determined by the chemistry of the crystallization medium depends both on the fungal metabolic activity and properties of underlying mineral substrate. The crystallization of calcium oxalates or carbonates was traced depending on the changing properties of the *A. niger* + *B. subtilis* co-culture medium. The stimulating effect of metals on acid production largely depends on metal concentration, nitrogen form in a medium, and other cultivation conditions. The biochemical potential of the *A. niger* makes it possible to consider this fungus as a promising object for bioremediation and other biotechnologies. The discussed original and published data contribute to the essential questions on patterns of oxalates formation within biofilms on different rocks and minerals; lay the foundations in the study of the crystal chemistry of microbial oxalates and outline new approaches to the development of nature-friendly biotechnologies, including the detoxification of heavy metals, using microscopic fungi, which are active producers of oxalic acid.

Chapter 2 - Filamentous fungi of the genus Aspergillus are potential sources of several enzymes and bioproducts. Aspergillus encompasses several distinct species, many of which possess promising properties and mechanisms for various applications. They are found naturally in diverse environments, exhibiting stability, selectivity, and easy assimilation of alternative carbon sources, making this fungus valuable for industrial purposes. *Aspergillus niger* has many properties that are industrially useful, including the ability to naturally accumulate compounds, and to produce biologically active secondary metabolites as well as many proteins and enzymes with biotechnological potential. Among the biotechnological and industrial enzymes produced by *Aspergillus niger* are amylases, cellulases, pectinases, peroxidases, lipases, xylanases, proteases, and laccases, used for pharmaceutical, textiles, food, beverages, and biorefineries. These biocatalysts often replace chemical hydrolysis processes or are used in the production of compounds such as peptides that have antibacterial, antifungal and antioxidant properties. They are also used in the production of products such as biosurfactants and fungal chitosan. This chapter will review recent advances in the prospecting of biocompounds and products with the most relevant biological and biocatalytic activity produced by *Aspergillus niger*. The challenges and future perspectives of biotechnological process extensions using *Aspergillus niger* will also be discussed.

Chapter 3 - *Aspergillus niger* is one of the most significant fungi that is comprehensively utilized in different natural fields. It has been utilized for a long time in the creation of extracellular enzymes and citric acid. Over the last two decades, *A. niger* has been created as a significant change host to overexpress nourishment enzymes. *A. niger* strains produce a progression of secondary metabolites by fermentation that has demonstrated to be an important biomaterial. From a natural perspective, it very well may be utilized as a bioadsorbent to detoxify and decolorize the wastewaters. In industrial and pharmaceutical applications, it can similarly be utilized as a cheap and attractive hotspot for business creation of some costly and helpful biopolymers, such as chitin, chitosan, and nourishment enhancements like chitin-glucan and glucosamine. This section is, for the most part, centred around the development of citric acid by *A. niger*.

In: *Aspergillus niger*
Editor: Eva Baughan

ISBN: 978-1-53618-080-0

Chapter 1

THE USE OF *ASPERGILLUS NIGER* IN MODELING OF MODERN MINERAL FORMATION IN LITHOBIOTIC SYSTEMS

Dmitry Yu. Vlasov[1,*], Olga V. Frank-Kamenetskaya[2], Marina S. Zelenskaya[1], Katerina V. Sazanova[3,4], Aleksei V. Rusakov[2] and Alina R. Izatulina[2]

[1]Department of Botany, St. Petersburg State University, St. Petersburg, Russia
[2]Department of Crystallography, St. Petersburg State University, St. Petersburg, Russia
[3]Komarov Botanical Institute RAS, St. Petersburg, Russia
[4]The Archive of the Russian Academy of Sciences Moscow, Russia

[*] Corresponding Author's Email: dmitry.vlasov@mail.ru.

ABSTRACT

The present review article summarizes the results of numerous research experiments on biomineralization on rocks and minerals surface carried out with the participation of fungus *Aspergillus niger*. The variations of fungus cultivation on liquid and solid media as well as in a moist chamber were reviewed. Metabolism of different strains of *A. niger* isolated from various habitats was discussed. The main attention was paid to the production of organic acids, especially oxalic acid, which plays a leading role in the interaction between fungus and rock. The influence of cultivation conditions on the metabolic activity of *A. niger* and of *A. niger* + *Bacillus subtilis* association was analyzed. It was revealed that among the components of culture medium that catalyse or inhibit the development and the metabolism of *A. niger*, we can, first of all, highlight the concentration of sugar, the chemical form of nitrogen, as well as the presence of calcium carbonate and heavy metals in the medium. Metabolism of *A. niger* and of *A. niger* + *B. subtilis* association in the same conditions was vastly different which significantly affects biomineralization.

The patterns of following oxalates formation on different mineral substrates by *A. niger* participation were revealed: calcium oxalates (weddellite and whewellite), strontium oxalates, copper oxalates (moolooite, fiemmeite, wheatleyite), iron oxalate (humboldtine), magnesium oxalate (glushinskite), manganese oxalates (lindbergite and falottaite) and finally lead and zinc oxalates which don't have mineral analogues yet met in nature – on cerussite and smitsonite. It was shown that morphogenetic patterns of oxalate formation determined by the chemistry of the crystallization medium depends both on the fungal metabolic activity and properties of underlying mineral substrate. The crystallization of calcium oxalates or carbonates was traced depending on the changing properties of the *A. niger* + *B. subtilis* co-culture medium.

The stimulating effect of metals on acid production largely depends on metal concentration, nitrogen form in a medium, and other cultivation conditions. The biochemical potential of the *A. niger* makes it possible to consider this fungus as a promising object for bioremediation and other biotechnologies.

The discussed original and published data contribute to the essential questions on patterns of oxalates formation within biofilms on different rocks and minerals; lay the foundations in the study of the crystal chemistry of microbial oxalates and outline new approaches to the development of nature-friendly biotechnologies, including the detoxification of heavy metals, using microscopic fungi, which are active producers of oxalic acid.

Keywords: *Aspergillus niger*, fungal-bacterial association, metabolic pathways, organic acids production, rocks and minerals, biomineralization, crystallization, oxalates, bioremediation

INTRODUCTION

In recent years, the world scientific community has shown significant interest in the mechanisms of biomineralization with the participation of microorganisms, which is associated with the study of modern mineral formation at the nano and micro levels, as well as processes and phenomena occurring at the border of living and nonliving. These fundamental results create the scientific basis for the development of modern nature-like biotechnologies in various fields of science and technology.

Fungus *Aspergillus niger* is one of the most popular objects of experimental research carried out to solve a wide range of fundamental and applied problems, including the study of patterns of microbial biomineralization, metabolic pathways, as well as biotechnologies aimed at bioremediation of polluted environments, industrial production of citric acid and others. Such widespread use of this fungus is due to its intense metabolic activity (including high acid production), ease of cultivation on nutrient media and the ability to adapt to different conditions.

The purpose of this multidisciplinary review is to summarize the results of numerous experiments on biomineralization in lithobiotic systems carried out with the participation of fungus *A. niger*. Special attention is paid to the peculiarities of the cultivation of the fungus in various media, the patterns of metabolism of its various strains, as well as the prospects for the use of *A. niger* in applied research.

1. The Cultivation of *Aspergillus niger* and *Aspergillus niger* + *Bacillus subtilis* Association in Different Conditions

One of the most effective tools for studying biological and biomineralogical processes occurring on solid natural substrates is their modeling under experimental conditions. And there are two main tasks that need to be solved experimentally. The first is related to studying the destruction of various rocks and minerals under the influence of aggressive microorganisms (microscopic fungi, bacteria and algae) under various conditions (Ehrlich 1999; Adamo and Violante 2000; Chen et al. 2000; Sterflinger 2000; Warscheid and Braams 2000; Salvadori and Municchia 2015). The second is related to the identification of the main factors and mechanisms of the formation of secondary minerals under the influence of microorganisms (Burford et al. 2003; Gadd 2004, 2007; Gadd et al. 2014).

Among micromycetes actively participating in the processes of transformation of stone substrates, much attention is paid to the fungus *A. niger*, which is wide spread and is an active destructor of natural and artificial stone. It is also known that stone can be affected by both individual species and communities, which is more common for natural conditions (Hoppert et al. 2004; Gorbushina 2007; Gadd 2017). Moreover, in the composition of the microbial communities (biofilms), combinations of fungi from the genus *Aspergillus* and bacteria of the genus *Bacillus* are often represented. These microorganisms were selected by us for the experiments.

The accumulated information on the modeling of biomineralization in laboratory conditions with the participation of *A. niger* (Table 1.1, 1.2) allowed us to experimentally reproduce the interaction models of this fungus and various types of natural stone in liquid and agarized media rich in nutrients, as well as in oligotrophic conditions of a moist chamber (Table 1.3). Based on the environmental characteristics of the studied microorganisms, we varied the conditions for the interaction of the fungus

A. niger, as well as the association of *A. niger* and *B. subtilis* with the mineral substrate.

A. niger (strain Ch4/07) was previously isolated from the damaged surface of Proconesos marble (Chersonesos, Crimea). The characterization of the strain was carried out in the Research Center for "Genomic technologies and cellular biology" of the All-Russian Research Institute of Agricultural Microbiology. The species identification of the strain was based on the sequence of the ITS region of rDNA (GenBank accession no. KF768341).

The bacterial strain *B. subtilis* 0.01 was isolated from the weathered rapakivi granite (Virolahti, Finland) and identified in the Institute of Molecular Genetics of the Russian Academy of Sciences by analysing the nucleotide sequence of the most part of the 16S rRNA gene. This strain is deposited in the RCAM collection under the number RCAMO4920.

Table 1.1. The results of the cultivation of *A. niger* in a liquid medium and a moist chamber on the surface of rocks and minerals

Cultivation conditions		Synthesized oxalates	Reference
Underlying mineral substrate	Culture medium		
Liquid culture medium			
Powder of Mn oxide nodules	AP1	Manganese oxalate dihydrate, weddellite, whewellite	Ferrier et al. 2019
Nickel chloride	Potato dextrose broth (PDB)	Nickel oxalate dihydrate	Magyarosy et al. 2002
Forsterite	Czapek medium, containing 10 g/L of glucose	Glushinskite	Cai et al. 2013
Chrysotile	Glucose solution	Glushinskite	Wilson et al. 1981
Zinc phosphate	Liquid nutrient medium*	Zinc oxalate	Sayer and Gadd 2001
Moist chamber			
Ivory boar tusk	Chapek-Dox (3-4 drops onto each surface)	Weddellite, whewellite	Pinzari et al. 2013

* Liquid culture medium contained, per l ddH_2O: 5g $(NH_4)SO_4$, 0.5g KH_2PO_4, 0.2g $MgSO_4{\cdot}7H_2O$, 0.05g $CaCl_2{\cdot}6H_2O$, 0.1g NaCl, 0.0025g $FeCl_3{\cdot}6H_2O$, 0.004g $ZnSO_4{\cdot}7H_2O$, 0.004g $MnSO_4{\cdot}4H_2O$, 0.0004g $CuSO_4{\cdot}5H_2O$, 20 g D-glucose.

Table 1.2. The results of the cultivation of *A. niger* on an agar medium with the addition of powders of natural and synthetic compounds

Agar medium composition		Synthesized oxalates	Reference
Malt Extract Agar (MEA), containing 0.5% (m/v) of biogenic manganese oxide birnessite		Manganese oxalate dihydrate (lindbergite)	Wei et al. 2012
Tap water agar (TWA), containing 20 g/L glucose +	Fluorapatite	Calcium oxalate	Adeyemi and Gadd 2005
	Galena	Lead oxalate	
Potato dextrose agar (PDA), containing lead carbonate or lead oxide		Lead oxalate	Sutjaritvorakul et al. 2015
Czapek-Dox medium, containing 0,2% (m/v) lead arsenate mimetite		Lead oxalate	Ceci et al. 2015
PDA, containing Pb foil*		Lead oxalate	Wu et al., 2020
PDA, containing synthetic PbS		Lead oxalate	Adeyemi 2009
MEA, containing 0.2% (m/v) lead lorophosphate - pyromorphite		Dihydrous and unhydrous lead oxalates	Sayer et al. 1999
MEA, containing 0.5% (m/v) zinc sulfide mineral ores		Zinc oxalate, weddellite	Wei et al. 2013
PDA, containing ZnO powder		Zinc oxalate	Sutjaritvorakul et al. 2013
MEA, containing cuprite		Copper oxalate	Sayer et al. 1997
MEA, containing rhodochrosis		Manganese oxalates	
MEA, containing gypsum		Calcium oxalate	Gharieb et al. 1998
MEA, containing strontium nitrate		Strontium oxalate	Sayer and Gadd 1997
Czapek-Dox medium, containing lantanium chloride heptahydrate		Lantanium decahydrate oxalate	Kang et al., 2019

* Fragments of Pb foil 1 cm × 0.5 cm ×0.5 mm.

Table 1.3. Characterization of model experiments conducted with the participation of *A. niger* using Czapek-Dox culture medium

Underlying stone substrate		Exposure (days)	pH	Synthesized oxalates
Name	Origin			
Liquid culture medium				
Marble	Carrara, Italy	13-45	5,0 – 5,5	
Dolomite	Quarry near Illesi village, Leningrad region, Russia	15-21	4,5 – 5,5	Weddellite, whewellite
Strontium carbonate synthetic powder	Produced by "Vecton" company	2-12	4,5 – 5,0	Strontium oxalate mono- and dihydrate
Hydroxyapatite powder	Synthesised by precipitation by M.A. Kuz'mina, department of crystallography, SPbSU	2-12		
Fluorapatite powder		2-12		
Natural fluorapatite crystal	Slyudyanka mineral deposit, Irkutsk region, Russia	2-21		
Horse tooth dentine	From alluvial deposits of the Late Pleistocene, Kursk region, Russia	2-21		
Mammoth tooth dentine	From alluvial deposits of the Pleistocene, Yakutia, Russia	2-21	4,5 – 5,5	Weddellite, whewellite
Liquid culture medium				
Southern elephant tooth dentine	From alluvial deposits of Neogene, Russia	2-21		Weddellite, whewellite
Fluorapatite ore	Kola Peninsula, Russia	5-12	4,0 – 5,0	Whewellite
Cuprite with brochantite	Gumeshevsky mine, Ural, Russia	3-21	3,5 – 4,5	Moolooite
Natural malachite		3-21		Moolooite, fiemmeite
Synthetic malachite	Synthesized (patent pending RU2159214)	3-21		Moolooite

Table 1.3. (Continued)

Underlying stone substrate		Exposure (days)	pH	Synthesized oxalates
Name	Origin			
Todorokite	Niki-Niki mnt., West Timor Island, Indonesia	2-30	4,5 – 3,0	Lindbergite, falottaite, weddellite, whewellite
Kutnohorite	Rhodopean, Bulgaria	2-30		
Magnesite	Satka, Ural, Russia	5-45	3,5 – 3,0	Glushinskite
Smitsonite	Central Asia	14-45	3,5 – 4,5	Zinc oxalate dihydrate
Cerussite	Central Scientific and Research Geological Museum of Academician F. N. Chernyshev (TSNIGR Museum), Federal State Budgetary Institution A. P. Karpinsky Russian Geological Research Institute (FGUP VSEGEI)	2-45	4,0 – 3,5	Anhydrous lead oxalate
Liquid culture medium				
Siderite	From FGUP VSEGEI collection	2-30	3,5 – 3,0	Humboldtine
Biotite on the surface of rapakivi granite	Montferrand Quarry, Virolahti, Finland	15-30	3,5 – 2,5	
Moist chamber				
Marble	Carrara, Italy	30-90	–	Weddellite
Dolomite	Quarry near Illesi village, Leningrad region, Russia	15-30	–	
Mammoth tooth dentine	From alluvial deposits of the Pleistocene, Yakutia, Russia	30-90	–	
Cuprite powder	Chemical synthesis	5-10	–	Moolooite
Malachite powder				
Brochantite powder				Moolooite, wheatleyite
Mammoth tooth dentine	From alluvial deposits of the Pleistocene, Yakutia, Russia	5-20	–	Weddellite, whewellite

Cultivation in a liquid and agar medium, as well as in a moist chamber, was carried out using the Czapek-Dox medium on the surface of blocks (1x1x0.5 cm) cut from various rocks and minerals, as well as powders of their synthetic analogues (Table 1.3, Figure 1.1). Czapek-Dox medium is a synthetic glucose-mineral medium with a well-known composition of mineral salts and sugars, optimal for the cultivation of *A. niger* and *A. niger* + *B. subtilis*: $NaNO_3$ – 3.0; KH_2PO_4 – 1.0; $MgSO_4 \cdot 7H_2O$ – 0.5; KCl – 0.5; $FeSO_4 \cdot 7H_2O$ – 0.015; glucose – 30.0 g/L (pH = 5.5). The choice of this medium allows to vary the concentrations of its components depending on the objectives of the experiment. It is also suitable for the analysis of metabolic products of fungi and bacteria formed during the experiment. For study the acid production by *A. niger* in addition to Czapek-Dox medium with nitrate form of nitrogen we also used Rollen medium with ammonium nitrate form (NH_4NO_3 – 3.0 g/l) and glucose 50.0 g/l to maintain the optimal C/N balance.

Samples of minerals and rocks were provided by the Museum of the Department of Mineralogy of St. Petersburg State University and the Central Scientific and Research Geological Museum of Academician F. N. Chernyshev (TSNIGR Museum).

1.1. *A. niger* Cultivation

1.1.1. In Liquid Culture Medium

In the Czapek-Dox liquid culture medium, mycogenous crystallization of calcium oxalates occurs rather quickly, which makes it possible to study the process of oxalate crystallization over time. In some studies, the culture filtrate of the *A. niger* fungus obtained on the Czapek-Dox liquid medium and AP1 was used to extract metals from the underlying substrates and to bind them during the formation of oxalates (Sayer and Gadd 2001; Yang et al. 2019, 2020).

In our experiments (Table 1.3, Figure 1.1) blocks cut from rock or mineral or a small amount of powder of the required composition were placed on the bottom of a Petri dish and 15 ml of Czapek-Dox liquid medium was added so that the surface of the underlying substrate was completely covered with the nutrient medium (Figure 1.1a). Inoculation was carried out with *A. niger* conidia and mycelium fragments from a 7-day-old fungal culture obtained on Czapek-Dox agar medium. The inoculum was placed on the surface of a liquid culture medium without submerging for better development of the surface culture of the fungus. The experiments were carried out at a temperature of +26°C. The duration of the experiments ranged from two days to five months.

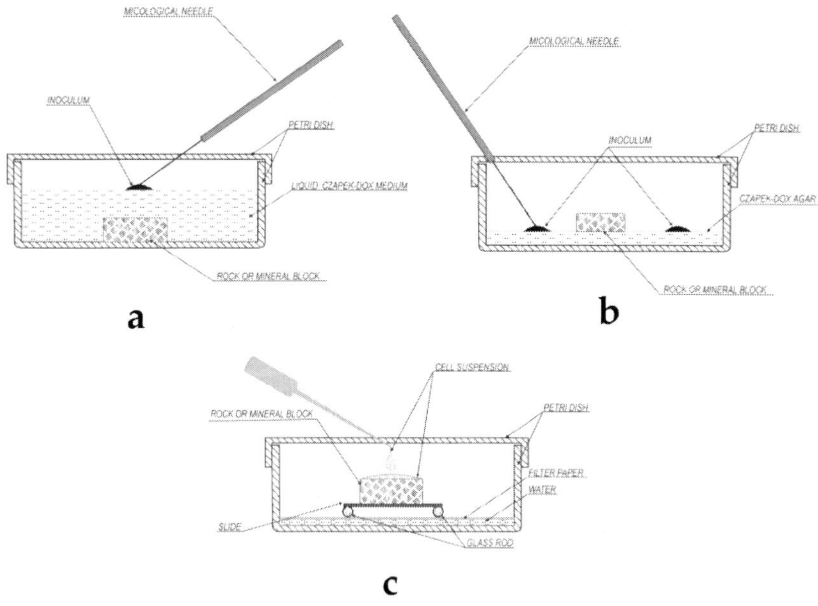

Figure 1.1. The scheme of experiments of biomineralization with the participation of the fungus *A. niger* and the association *A. niger* + *B. subtilis*: a – in a liquid nutrient medium; b – on an solid medium; c – in a moist chamber.

1.1.2. On Solid Culture Media

Studies on the formation of oxalates under the influence of *A. niger* were carried out using mostly various solid culture media (Table 1.2). Most often, Malt extact agar (MEA) medium and potato-glucose medium were used. Crushed rock fragments and minerals, as well as powders of synthetic compounds were applied directly to the nutrient medium as well.

In our experiments, the corresponding inorganic substrates were placed on the surface of the Czapek-Dox agar medium in Petri dishes. The inoculum of *A. niger* fungus was introduced with a mycological needle onto the medium at a distance of 1-2 cm from the substrate (Figure 1.1.b, Table 1.3). The incubation period was 3 weeks at a temperature of +25 °C.

1.1.3. In Moist Chamber

Oligotrophic conditions that closely match natural conditions can be simulated in moist chamber. In Pinzari et al. (2013) (Table 1.1) the authors placed fragments of animal bones on whatman paper discs previously put in Petri dishes with a diameter of 15 cm. After that a spore suspension was prepared from a 7-day *A. niger* culture in distilled water and mixed with Czapek-Dox liquid nutrient medium. The resulting mixture in a quantity of 3-4 drops was applied to the surface of bone fragments. Bone samples onto which a mixture of Czapek-Dox nutrient medium and distilled water containing no fungal spores was applied were used as control. Petri dishes with inoculated bones were placed in double bottom glass vessels into which distilled water was poured to maintain 100% humidity. Incubation was carried out for 10 days.

In our experiments in moist chamber (Figure 1.1c, Table 1.3.), we used a conidia suspension prepared on the basis of the Czapek-Dox liquid medium. *A. niger* spore suspension, the concentration of which was 10^4-10^5 spores per 1 mL, was applied in an amount of 0.1 mL on the surface of a rock fragment or mineral. Thus, the Czapek-Dox medium landed the surface of the stone at the time of inoculation together with a suspension of spores.

1.2. *A. niger* + *B. subtilis* Cultivation

In known biomineralization experiments, particular types of microorganisms have usually been used. There are practically no works where biomineralization was studied under the influence of microbial associations, for example, of fungus and bacteria. It is known that in microbial associations (biofilms) microorganisms are able to exert a mutual influence on the processes of growth, metabolism, and adaptation to external conditions (Benoit et al. 2015). The authors noted that bacteria of the genus *Bacillus* are able to interact with fungal hyphae, participating in the formation of biofilms.

1.2.1. In Liquid Medium

In the experiments with liquid medium marble blocks were put on the bottom of Petri dishes and 15 ml liquid Czapek-Dox medium was added so that the surface of blocks was covered. The glucose concentration in the medium was 1, 5, 10, 30 g/L (Table 1.4). The inoculation was carried out by the suspension of bacterial cells and fungal conidia. The number of bacterial cells in 1 mL of the initial suspension was 10^8, fungal conidia – 10^4, which corresponds to the concentrations of cells and conidia on the destroyed limestone (Somova et al. 1998; Kurakov et al. 1999). The cultivation time was: 14, 21, 30 and 60 days.

Table 1.4. The conditions of model experiments carried out with the participation of *A. niger* and *A. niger* + *B. subtilis*

Organisms	Glucose concentration, g/L	Days of growth		
		14	21	30
		pH		
B. subtilis	1	7.5	7.8	8.2
	5	7.5	8.2	8.6
	10	7.0	7,4	8.0
	30	6.5	7.2	8.1
A. niger + *B. subtilis*	1	7.5	8.0	8.0
	10	4.5	5.0	7.0
	30	3.0	5.5	5.0

1.2.2. In Moist Chamber

In experiments on co-cultivation of *A. niger* and *B. subtilis*, a suspension of fungal spores and bacterial cells in the Czapek-Dox nutrient medium was simultaneously applied to the surface of the substrate. The *A. niger* spore concentration was 10^4–10^5, and the concentration of *B.*

subtilis cells was 10^7–10^8 per 1 mL of suspension. Microorganism cell suspension (0.1 mL) was applied on the marble block surface. Nutrient medium was not added later and distilled water was added to the filter paper on the bottom of the Petri dish every two weeks. Thus, the conditions created for the growth of microorganisms on a mineral substrate included high humidity and a minimum amount of nutrients.

All experiments were in triplicate. For comparison, control experiments were carried out in parallel without the participation of fungi and bacteria. Precipitates formed during the experiment and containing crystallization products and microbial biofilms were investigated directly on the underlying inorganic substrate. The phase composition of the precipitates was determined by means of X-ray powder diffraction; morphology of synthesized crystals - by optical and (mainly) scanning electron microscopy (SEM) methods. Scanning electron microscopy with elemental composition control (EDXS) was also used to identify phases by morphological characteristics in the case of the diagnosis of secondary calcite on the surface of calcite marble, as well as oxalates and other phases formed, if the substance for X-ray analysis was insufficient.

In order for cultivation conditions to be closer to natural lithobiotic systems, in the future it is necessary to proceed to establish the experiments with associations of microorganisms typical for natural biofilms.

2. Metabolism *Aspergillus niger* and *A. niger* + *Bacillus subtilis* Association

2.1. *A. niger* Metabolism

2.1.1. The Main Metabolites and Their Transformations

More than 800 unique secondary metabolites of the fungi genus *Aspergillus* are known to date, 523 of which are included in chemical databases such as PubChem, ChemSpider, ChEBI, TOXNET. Among those secondary metabolites, 213 were isolated from *A. flavus*, *A. fumigatus*, *A. oryzae*, and *A. niger*, and 90 were isolated precisely from *A. niger* (Vadlapudi et al. 2017).

The exometabolics of *A. niger* include about 500 metabolites, and the main ones are alcohols, aldehydes, esters, hydrocarbons, ketones and terpenoids (Costa et al. 2014).

The volatile fraction of *A. niger* metabolites includes 428 compounds, among which the main ones are hydrocarbons, alcohols, esters, ketones and aldehydes. These metabolites can be associated with various metabolic pathways, such as amino acid metabolism, biosynthesis and metabolism of fatty acids, the degradation of aromatic compounds, the synthesis of mono- and sesquiterpenoids, and the breakdown of carotenoids. The structure of *A. niger* molecular biomarkers was established, among which there were 44 metabolites including 16 metabolites that play a major role for distinguishing between fungi (Costa et al. 2016).

The best known biochemical property of *A. niger* is its ability to excrete low molecular weight organic acids (gluconic, citric, malic, fumaric, succinic and oxalic) (Max et al. 2010; Joseph et al. 2015; Sazanova et al. 2015, 2016; Show et al. 2015; Almousa et al. 2018). The ability to actively excrete citric acid was the reason for *A. niger* usage in industrial acid production (Andersen et al. 2011; Show et al. 2015; Almousa et al. 2018; Ozdal and Kurbanoglu 2019).

Unlike other acids excreted by *A. niger*, gluconic acid is a product synthesized mainly extracellularly. At the first stage, glucose is oxidized by the glucose oxidase enzyme to D-glucono-1.5-lactone. Hydrolysis of lactone to gluconic acid can occur spontaneously in an aqueous solution, however, the rate of this reaction increases significantly under the action of the gluconolactonase enzyme. *A. niger* has a gene encoding intracellular glucose oxidase as well (Pel et al. 2007). Based on our recent research, *A. niger* converts up to 25% of the glucose in the medium into gluconic acid. In this case, apparently, the active oxidation of glucose to gluconate promotes more efficient use of the substrate and, accordingly, more active fungal growth (Sazanova et al. 2016).

The formation of citric, succinic, fumaric and malic acids occurs as a result of the oxidation of sugars inside the cell. With a sufficient amount of oxygen, the pyruvic acid molecule formed as a result of glycolysis is decarboxylated to form acetyl-CoA by the mitochondrial pyruvate dehydrogenase complex.

Acetyl-CoA interacts with oxaloacetate to form citric acid (Soldatenkov 1971; Kretovich 1980; Peksel et al. 2002). The citrate synthase enzyme that catalyzes this reaction is inhibited by Mg^{2+} and adenosine-triphosphoric acid (ATP), but not by citric acid (Papagianni 2007).

Another way for the metabolism of pyruvic acid is possible as well. In the 1960s CO_2 fixation by the pyruvate carboxylase enzyme with the formation of oxalacetic acid in fungi was proved (Woronick and Johnson 1960; Bloom and Johnson 1962). Later, one mitochondrial and one cytoplasmic pyruvate carboxylase enzyme were found in *A. niger* (Pel et al. 2007). Oxaloacetic acid formed in this way, if synthesized in the cytoplasm, can be transported to mitochondria (via an oxaloacetate translocator in exchange for malate) and incorporated into the tricarboxylic acid cycle. In addition, oxaloacetate can be converted to malic acid by cytoplasmic or mitochondrial malate dehydrogenase (Ma et al. 1981; Pel et al. 2007).

The metabolism of citric acid formed during the citrate synthase reaction can take several paths. In mitochondria, citric acid takes major role in subsequent reactions of the Krebs cycle. In addition, citrate may be split into oxaloacetate and acetate. This reaction is carried out by the citrate lyase enzyme in the presence of Mg^{2+} or Mn^{2+} ions. A significant portion of citric acid is transported through the citrate/malate transporter from mitochondria to the cytosol. *A. niger*, like other fungi, contains two ATP-citrate lyases (cytoplasmic and mitochondrial). ATP-citrate lyase catalyzes the reaction: citrate + CoA + ATP $\leftrightarrow$ acetyl-CoA + oxaloacetate + ADR + P*i* (Pel et al. 2007).

The next reaction of citric acid conversion in the tricarboxylic acid cycle is the reaction of isomerization of citrate to isocytrate under the action of the aconitase enzyme. Isocitric acid is decarboxylated to α-ketoglutarate by a NAD-dependent isocitrate dehydrogenase and is oxidatively decarboxylated to form succinyl-CoA. The conversion of succinyl-CoA to succinic acid is catalyzed by the succinyl-CoA synthase enzyme and is accompanied by the formation of ATP. Generally, succinic acid does not accumulate in large amounts in the extracellular solution. Its amount is usually not more than 25% of the total amount of acids produced by the fungi (Bercovitz et al. 1990). In addition to the oxidation of α-ketoglutarate, succinic acid can also be formed as a result of fumaric acid reduction by succinate dehydrogenase and from isocitrate during the glyoxylate pathway (Munir et al. 2001; Magnuson and Lasure 2004).

The oxidation of succinic acid to fumaric acid is carried out with the participation of succinate dehydrogenase. The cytoplasmic pathway for the synthesis of fumaric acid from malic acid under the action of fumarate hydratase (fumarase) present in the cytoplasm is also possible (Carol et al. 2008).

The formation of malic acid, just as fumaric acid, can occur in mitochondria or in the cytoplasm. In the tricarboxylic acid cycle, fumaric acid under the action of the fumarase enzyme attaches water and turns into malic acid. Further, malic acid under the action of the malate dehydrogenase enzyme is oxidized to oxaloacetic acid with the formation of NAD (Sodatenkov 1971; Peksel et al. 2002). In fungi, the presence of three cytoplasmic and one mitochondrial malate dehydrogenas enzymes is assumed (Pel et al. 2007). Cytoplasmic malate dehydrogenase generates malic acid from oxaloacetic acid, which is synthesized as a result of CO_2 fixation by cytoplasmic pyruvate carboxylase (Ma et al. 1981; Magnuson and Lasure 2004). Further, malate formed in the cytoplasm can be transported to mitochondria, oxidized to oxaloacetate, and condensed with acetyl-CoA with the formation of citric acid (Kretovich 1980; Papagianni 2007).

It is assumed that the excretion of di- and tricarboxylic acids of the Krebs cycle occurs with the participation of transport proteins (Gallmetzer and Burgstaller 2002; Aliverdieva et al. 2010). The transporting of most acids (malate, succinate, fumarate, oxaloacetate, malonate) occurs, apparently, by symporting its anion with a proton (Jennings 2007;

Aliverdieva et al. 2010). Organic acid transporters are present both on the plasmalemma and on the mitochondrial membrane (smaller in size for the latter). The transporters on plasmalemma have 11 or 12 hydrophilic transmembrane α-helical segments forming a channel, while mitochondrial transporters have 6 segments (Aliverdieva et al. 2010). The transport systems of malic and succinic acids in most fungi are suppressed by glucose (Jennings 2007).

Oxalic acid plays a key role during mineral formation. The mechanism of oxalic acid formation from oxaloacetate in *A. niger* was first proposed in 1956 by Hayaishi et al., who associated the formation of this acid with the activity of the oxaloacetate acetylhydrolase enzyme (oxaloacetase) (Hayaishi et al. 1956). Since the inhibition of the of tricarboxylic acids cycle in mitochondria did not cause a decrease in the formation of oxalic acid, the cytoplasmic localization of oxaloacetase was assumed and was later confirmed by fractionation of *A. niger* mycelium cells (Kubicek et al. 1988).

The hydrolysis of oxaloacetic acid by the cytoplasmic oxaloacetate acetylhydrolase was accepted as the main method for the synthesis of oxalic acid (Munir et al. 2001; Magnuson and Lasure 2004; Pel et al. 2007). The subsequent studies showed that strains of *A. niger* with insufficient oxaloacetase activity did not produce oxalic acid (Ruijter et al. 1996). The experiments of Pedersen (2000) and Han et al. (2007) demonstrated that the destruction of the *oah* gene encoding oxaloacetase in *A. niger* leads to the formation of mutants that are unable to produce oxalic acid. It was suggested that the presence of this gene in fungi is necessary for the biosynthesis of oxalate (Pedersen et al. 2000; Han et al. 2007).

2.1.2. A Change of A. niger Metabolism under Various Conditions of Cultivation

The intensity of excretion and the quantitative ratio of the organic acids produced by fungi varies significantly depending on growth conditions, the stage of culture development, and strain characteristics of the fungus.

One of the main factors affecting the quantitative composition of excreted acids is the *composition of nutrient medium*, primarily the sources of carbon and nitrogen. The best carbon sources for acid formation by *A. niger* are glucose or sucrose. From the industrial standpoint using of molasses (Khattab et al. 2017) and sucrose (Hossain et al. 1984) is most effective for citric acid production by fungi, while glucose and fructose are less suitable for the production. Lactose is even less favorable for the biosynthesis of citric acid. When *A. niger* is cultured on galactose, citric acid does not form at all (Hossain et al. 1984). The productivity of oxalic acid by *A. niger* practically does not change when glucose, fructose, sucrose, gluconate, xylose, acetate, and glycerol are used as a carbon source (Ruijter et al. 1999). Gluconic acid is produced in greatest quantities using glucose (Singh et al., 2001), although there is evidence that gluconic acid is often formed by fungi using other carbon sources (Ramachandran et al. 2006). In most cases, the formation of acids by fungi increases with an increase in the concentration of carbohydrates in the medium (Braams 1992; Gallmetzer and Burgstaller 2002; Rosling et al. 2004; Magnuson and Lasure 2004; Al-Shehri and Mostafa 2006).

According to our data (Sazanova et al. 2016), citric, malic, succinic, and fumaric acids are released only at high sugar concentrations (more than 3%). In this case, not only the concentration of sugars, but also the form of nitrogen in the medium is important. In these experiments for fungi cultivation, two culture media differing in the nitrogen source were used: Czapek medium ($NaNO_3$ - 3.0 g/L) and Rollen medium (NH_4NO_3 - 3.0 g/L). Glucose content was 30.0 g/L in Czapek media and 50.0 g/L in Rollen media to maintain the optimal C/N balance.

Table 2.1. The formation of *A. niger* organic acids in liquid media with different nitrogen sources on the 10th day of cultivation

Medium (nitrogen source)	Organic acid amounts, mg/g of micelium				
	Oxalic	Citric	Succinic	Malic	Fumaric
Czapek ($NaNO_3$)	57.8 ± 7.8	5.5 ± 1.2	1.0 ± 0.2	-	-
Rollen (NH_4NO_3)	0.11 ± 0.02	6.6 ± 1.2	0.28 ± 0.03	0.45 ± 0.9	0.04 ± 0.04

The release of oxalic acid is less dependent on the concentration of sugars than the release of acids of the Krebs cycle, and it decreases when medium contains an ammonium source of nitrogen compared to nitrate (Table 2.1.) (Sazanova et al. 2016). On media with the same concentration of carbohydrates the amount of excreted acids of the Krebs cycle is also higher when using nitrate as the only nitrogen source than when using NH_4NO_3 (Sazanova et al. 2015). One of the probable reasons for the lower production of acids on ammonium and nitrate-ammonium media, compared to nitrate, is most likely associated with a greater need for carbon skeletons for the disposal of ammonium toxic for microorganisms. Ammonia is believed to be a more easily assimilable form of nitrogen for fungi (Carlile et al. 2001; Meti et al. 2011). However, it is known that NH_4^+, unlike nitrate, has a toxic effect. For its utilization, fungi require continuous synthesis of carbon skeletons (Britto and Kronzucher 2002; Hess et al. 2006). To increase the synthesis of pentoses, pentose phosphate pathway becomes more intense when ammonium is used as a source of nitrogen (Holligan and Jennings 1972).

It is known that some organic acids excreted by fungi can be absorbed by mycelium and used as a source of carbon (Müller 1986; Elshafei 1989; Ghorbani et al. 2007). It was also suggested that after the complete depletion of carbohydrates in a nutrient medium, previously excreted acids are gradually absorbed by mycelium and processed into other products (Ghorbani et al. 2007). Perhaps the intense oxidation of sugars into acids contributes to easier assimilation of sugars.

Several sources indicate the relationship between the accumulation of acids in the medium and the stage of ontogenesis. Some authors note that the production of citric acid occurs most intensively in the early days of growth, and then slows down (Abín et al. 2002; Betlej and Grąz 2006). According to other data, the greatest amount of citric acid accumulates at the end of the log phase and at the beginning of the stationary growth phase (Ali et al. 2002a).

Our analysis of the culture fluid carried out on the 3rd, 7th, and 10th day of *A. niger* growth revealed a number of features of organic acids release depending on the age of the culture (Table 2.2). At the initial stage of growth (3rd day), mainly gluconic acid was present in the culture fluid, along with citric and oxalic acids in small amounts. With further cultivation, the amount of gluconic acid in *A. niger* culture fluid decreased. In this case, active production of oxalic acid occurred, the amount of citric acid increased, and malic, fumaric and succinic acids also appeared. Thus, at the early stages of ontogenesis, active formation of gluconic acid occurs. The largest amount of citric, malic, fumaric and succinic acids is recorded at the end of the log phase of mycelial growth. In the stationary phase of growth, oxalic acid accumulates in the medium, and the amount of other organic acids decreases (Sazanova et al. 2016).

The acidity of the medium is a key parameter of the chemical selectivity of the biosynthesis of organic acid formation by *A. niger*. Low acidity promotes the production of mainly oxalic (Ruijter et al. 1999; Carlile et al. 2001; Ali et al. 2002a; Al-Shehri and Mostafa 2006) and gluconic acid (Braams 1992). The activity of the oxaloacetate hydrolase enzyme, responsible for the decomposition of oxaloacetate to oxalate and acetate inside the fungal cell, is highest at pH = 6 (Walaszczyk et al. 2018).

Table 2.2. Organic acids formed by *A. niger* during ontogenesis

Day of growth	Organic acids, mg/g of micelium					
	Gluconic	Citric	Malic	Fumaric	Succinic	Oxalic
3	167.5 ± 18.2	2.1 ± 0.5	-	-	-	40.7 ± 6.6
7	37.2 ± 8.2	15.4 ± 6.5	0.1 ± 0.2	0.1 ± 0.3	0.1 ± 0.3	344.4 ± 40.4
10	9.8 ± 2.5	5.0 ± 0.7	-	-	-	380.6 ± 35.2

The optimal value for the biosynthesis of gluconic acid is pH = 6.5. When the pH of the nutrient solution is less than 3.5, the activity of the glucose oxidase enzyme is completely deactivated (Anastassiadis and Rehm 2006). The acidity of the extracellular solution is one of the main regulators of the activity of oxaloacetase. With a decrease in pH, the activity of this enzyme is suppressed (Kubicek et al. 1988; Ruijter et al. 1999). High acidity is believed to induce the formation of citric acid (Carlile et al. 2001; Zhang 2002a, 2002b). It is possible that the formation of citrate is enhanced at low acidity due to a decrease in carbon costs for the formation of oxalic and gluconic acids (Ruijter et al. 1999). Mutants of *A. niger* lacking glucose oxidase (*gox C*) and oxaloacetase (*prt F*) produced citric acid in large quantities and at neutral pH (Zhang et al. 2007).

A significant influence on the ability of fungi to produce organic acids is exerted by elements coming from the underlying stone substrate and the environment. Even in the early works on the study of the ability of fungi to produce organic acids (Foster 1950; Butkevich 1957), it was noted that calcium significantly affects the formation of organic acids by fungi and especially the oxalic acid. The addition of calcium chloride to the medium can stimulate the release of citric acid in *A. niger* (Pera and Callieri 1999). Intensive production of oxalic acid is often observed in fungi that live on substrates with a high content of Ca^{2+} (Adamo and Violante 2000; Kolo and Claeys 2005). The production of oxalic acid also significantly increases the deficiency of magnesium and potassium (Paris et al. 1996).

According to our experiments (Barinova et al. 2010), the presence of calcium carbonate in the medium promotes the release of primarily oxalic acid by fungi, which content in the medium was doubled compared to the case without the addition of $CaCO_3$. At the same time, the amount of citric acid excreted by *A. niger*, on the contrary, turned out to be 40% lower than on a medium without $CaCO_3$. Succinic, malic, and fumaric acids were present only on a medium not containing calcium carbonate, and were present in amounts less than 0.2 mg/mL. Low concentrations of minor acids in a medium with calcium carbonate indicates certain changes in the processes of organic acid metabolism under the influence of $CaCO_3$, associated with its stimulating effect on oxalic acid biosynthesis and, accordingly, an increase in carbon requirement for its formation.

The effect of $CaCO_3$ can be associated with the presence of Ca^{2+} ions, which neutralize the formed acids, increase the carbon dioxide content in the medium, and also alkalize the extracellular solution (Kubicek et al. 1988; Ruijter et al. 1999; Pedersen et al. 2000; Anastassiadis and Rehm 2006). We conducted an experiment comparing media with and without calcium (Barinova et al. 2010). Micromycetes were cultivated on liquid nutrient media: Czapek-Dox medium (pH 5.8) – medium I, Czapek-Dox medium + 0.2% $CaCO_3$ (pH 6.8) – medium II, Chapek-Dox medium + NaOH (pH 6,8) – medium III, Czapek-Dox medium + Na_2CO_3 (pH 6.8) – medium IV. These media were selected to create variance in the presence of Ca^{2+} and CO_3^{2-} ions (separately and together), as well as their absence in the medium. The results of the experiment showed that the release of oxalic acid in the presence of calcium carbonate more than doubled (Figure 2.1). In the presence of Na-containing compounds, the amount of oxalate increased to 60% in *A. niger*.

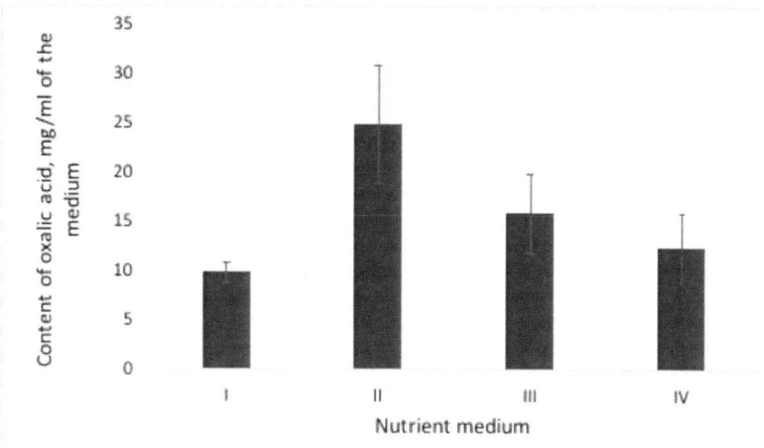

Figure 2.1. The amount of oxalic acid excreted by *A. niger* on the 10th day of growth on Czapek-Dox medium (pH 5.8) (I), and media with various additives: 0.2% $CaCO_3$ (pH 6.8) (II), NaOH (pH 6.8) (III), Na_2CO_3 (pH 6.8) (IV).

Among all the trace elements Zn, Mn, Fe, and Cu have the most significant effect on the metabolism of organic acids in fungi (Ali et al. 2002b; Papagianni 2007). Many studies demonstrated that the addition of certain metals to a culture medium leads to the activation of the formation of organic acids by fungi (Gibson and Mithell 2004; Fomina et al. 2005; Munir et al. 2005; Betlej and Grąz 2006). Depending on the concentration, metals can also suppress the formation of organic acids (Sazanova et al. 2015). An increase in the concentration of copper in the medium stimulates the synthesis of oxalic, citric, and some other acids by fungi (Gibson and Mithell 2004; Fomina et al. 2005; Munir et al. 2005; Betlej and Grąz

2006). Zinc causes a sharp increase of citric acid synthesis and also contributes to the synthesis of malic acid (Fomina et al. 2005). A number of authors have demonstrated that an increase in the concentration of some other metals (Cd, Ni, Pb) in the medium stimulates the synthesis of oxalic, citric, malic, and some other acids by fungi (Gibson and Mithell 2004; Fomina et al. 2005).

On the other hand, it is the lack of certain trace elements that contributes to the production of citric acid by fungi (Papagianni 2007). It is possible that an active fungal release of acids that are intermediates for the oxidation of sugars on carbohydrate rich and trace element poor media occurs due to impaired functioning of enzyme systems. The lack of iron ions, the cofactor of aconitase, increases the formation of citric acid by blocking the subsequent reaction in the Krebs cycle (Braams 1992; Carlile et al. 2001; Odoni et al. 2017). According to Odoni et al. (2017), the amount of iron added to the culture medium is a key factor for the excretion of citric acid by *A. niger*. Manganese has a stimulating effect on oxalic acid synthesis, since it is necessary for the functioning of oxaloacetase (Hayaishi et al. 1956; Lenz et al. 1976). In addition, Mn activates isocitrate dehydrogenase (Williams and Pittman 2010). The addition of manganese to the medium at concentrations of $1x10^{-6}$ - $4x10^{-6}$ leads to an increase in the production of citric acid in some strains and to a decrease in its production in others (Ali et al. 2002b). It is also known that a lack of aluminum has a stimulating effect on the production of organic acids by fungi (Lapeyrie et al. 1991). The addition of manganese to the medium at concentrations of $1 \cdot 10^{-6}$ - $4 \cdot 10^{-6}$ leads to an increase in citric acid production by some strains and to a decrease in its production by several others (Ali et al. 2002b). It is also known that a lack of aluminum has a stimulating effect on the production of organic acids by fungi (Lapeyrie et al. 1991).

2.1.3. Strains Features of A. niger Metabolism

The variability fungal strains may be due to their habitat conditions, and thus it can indicate specific properties of the substrate, it can also depict climatic and environmental conditions of the region.

Our comparative analysis of acid production activity of ten *A. niger* strains isolated from the surface of natural and artificial substrates (plaster, natural stone, wood, paper and cardboard) from different regions (Antarctica, Egypt, Sevastopol, St. Petersburg) showed that acid formation and excretion rates varies greatly depending on the type of substrate from which the strain was obtained (Sazanova and Vlasov 2020). The results of the study, presented in Table 2.3., indicate that the strains isolated from the surface of plaster had the greatest differences in the ability to excrete oxalic acid. The weakest producers of oxalate were *A. niger* strains isolated from the surface of paper, cardboard and wood. Strains isolated from the surface of natural stone excreted 40-50% less oxalic acid than strains isolated from the plaster coating, but significantly more than strains isolated from the surface of the paper. At the same time, one of the two strains isolated from paper possessed the greatest ability to produce citric acid. The plaster coating is a calcareous material used in construction. As mentioned above, the addition of calcium carbonate to the medium enhances the production of oxalic acid (Tuason and Arocena 2009; Barinova et al. 2010). Fungi and plants growing in calcium-rich soils can use organic acids to bind its excess amounts (Adamo and Violante 2000; Tuason and Arocena 2009). It can be assumed that the overproduction of oxalate is a physiological property of fungi under the conditions of their habitat on stony (both natural and artificial) substrates. On other types of substrate, such as wood and paper, oxalic acid can contribute to their destruction, but probably does not have an adaptive value for the fungus.

2.2. *A. niger* + *B. subtilis* Association Metabolism

The interaction between microorganisms in a community affects their growth and metabolism. The metabolic products of some species of fungi and bacteria can both suppress and activate the biochemical processes of other species, and also be used by them as a source of nutrient (Boonchan 2000; Gilbert et al. 2002; Berdoulay et al. 2009, Keshari et al. 2013). It is known, for example, that the bacteria *Salmonella typhimurium* and *A. niger* stably interact for several weeks without a noticeable negative effect. Bacterial growth is stimulated after interaction is established. In addition, bacterial biofilms protect the fungus from external influences. Thus, the *Salmonella-Aspergillus* interaction can be defined as mutualistic (Balbontín et al. 2014).

Recent studies have shown that *B. subtilis* bacteria attach to the root of a plant and form complex communities in the rhizosphere. When grown in the presence of *A. niger*, bacteria similarly interact with the fungus attaching and then growing on hyphae. Both fungi and bacteria are thought to alter their metabolism during this interaction. Transcription of genes associated with the antifungal and the supposed antibacterial defense mechanism of *B. subtilis* and *A. niger*, respectively, decreases when bacteria attach to mycelium (Benoit et al. 2015).

Extracellular polymeric substances (EPS) produced by bacteria including *B. subtilis* can play a significant role in microbial interactions as well as contribute to the formation of secondary calcite (Ercole et al. 2007; Dhami et al. 2013; Zhu and Dittrich 2016). EPS are used as nuclei for reprecipitation of calcite crystals from solution (Cao et al. 2016). EPS of *B. subtilis* includes a wide variety of molecules: polysaccharides, lipopolysaccharides, glycolipids, lipids and proteins (Marvasi et al. 2010). In general, data on the ability of bacteria to participate in the crystallization of oxalates is very limited. Besides, there is not much data about the acidification activity of bacteria *Bacillus* spp. (Bonaventura et al. 1999). More studies are devoted to the oxalotrophic activity of *B. subtilis* (Tanner and Bornemann 2000; Schneider et al. 2012; Lee et al. 2014).

The analysis of *B. subtilis* liquid culture carried out by us revealed that extracellular low weight carboxylic acids produced by bacteria include 2-hydroxypropanoic acid, oxalic and succinic acids (Sazanova et al. 2020). Oxalic acid, directly involved in the formation of calcium oxalate, was found in the medium on the 21st day of growth. Amount of oxalic acid was approximately the same at different glucose concentrations and slightly increased with time.

Table 2.3. The formation of organic acids by various strains of *A. niger* (10th day of growth)

Strain	Place of isolation	Substrate	Year of isolation	Organic acids (mg/g of micelium)					
				Oxalic	Succinic	Fumaric	Malic	Citric	Gluconic
Ch 4/07	Sevastopol	Marble	2007	246.3 ± 28.1	0.9 ± 0.1	0.4 ± 0.1	0.9 ± 0.3	1.7 ± 0.6	5.6 ± 2.4
S 1/11	Saint Petersburg	Plaster	2011	512.5 ± 78.6	0.5 ± 0.7	0.40 ± 0.06	0.8 ± 0.2	27.3 ± 58.3	-
P 2/08	Saint Petersburg	Cardboard	2008	122.6 ± 12,5	0.4 ± 0.04	0.23 ± 0.04	0.16 ± 0.03	0.21 ± 0.04	-
A 1/10	Antarctica	Rocky substrate	2010	201.5 ± 38.0	0.8 ± 0.2	0.36 ± 0.06	0.24 ± 0.07	7.4 ± 0.9	6.6 ± 0.8
A 2/12	Antarctica	Gneiss	2012	255.2 ± 34.1	0.7 ± 0.2	0.5 ± 0.1	0.7 ± 0.1	8.1 ± 1.1	7.2 ± 0.4
E 1/04	Egypt	Marble	2004	140.2 ± 12.4	-	-	-	1.8 ± 0.4	-
S 2/12	Saint Petersburg	Plaster	2012	516.0 ± 39.1	-	-	-	-	4.4 ± 0.9
P 1/12	Saint Petersburg	Paper	2012	140.4 ± 83.1	-	-	-	34.2 ± 0.6	7.1 ± 2.4
B 1/10	Saint Petersburg	Wood	2010	143.8 ± 23.1	0.5 ± 0.1	0.3 ± 0.1	0.8 ± 0.2	5.4 ± 1.0	13.2 ± 3.1
Bm-53	Saint Petersburg	Paper	2010	133.3 ± 32.8	-	-	-	-	-

Table 2.4. Metabolic products of microorganisms and pH values of cultural liquids at various glucose concentrations

Organism	Glucose concentration, g/l	Days of growth						
		14		21			30	
		pH	Oxalic acid, mcg/mL	pH	Oxalic acid, mcg/mL	EPS, mcg/mL	pH	Oxalic acid, mcg/mL
Bacillus subtilis	1	7.5	-	7.8	1.0 ± 0.4	328 ± 44	8.2	1.1 ± 0.3
	5	7.5	-	8.2	0.6 ± 0.4	405 ± 58	8.6	1.0 ± 0.3
	10	7.0	-	7,4	1.1 ± 0.3	1121 ± 181	8.0	1.5 ± 0.3
	30	6.5	-	7.2	0.8 ± 0.1	1460 ± 204	8.1	3.4 ± 0.4
Bacillus subtilis + *Aspergillus niger*	1	7.5	2.2 ± 0.5	8.0	2.0 ± 0.1	437 ± 66	8.0	15.5 ± 2.0
	10	4.5	153.8 ± 13.4	5.0	1032.9 ± 89.1	1.1 ± 0.2	7.0	1740.0 ± 99.4
	30	3.0	740.0 ± 20.8	5.5	3275.3 ± 232.1	10 ± 2	5.0	5275.1 ± 100.4

A mutual influence between the metabolism characteristics of *A. niger* fungus and *B. subtilis* bacteria can be observed within their association. According to our data total amount of EPS in the culture of *B. subtilis* increased with the increase of sugar concentration in the medium (Table 2.4). In the co-culture of *B. subtilis* + *A. niger*, the amount of EPS was similar to the amount of EPS in the monoculture of bacteria only at low glucose concentrations (0.1%). At higher glucose concentrations the active growth of fungus inhibited the growth of bacteria and the EPS content was very low. The metabolic activity of micromycetes can suppress the formation of bacterial EPS. When glucose concentrations are high enough the acid producing activity of the fungus increases, and possibly the acid producing activity of the bacteria, which leads to lowering of pH of the medium, which in turn prevents the formation of bacterial EPS. At low glucose concentrations, the acid producing activity of fungi does not significantly affect the formation of bacterial EPS.

The production of organic acids in the co-culture of *B. subtilis* and *A. niger* was also significantly dependent on glucose concentration. Only oxalic acid was released in medium at a low glucose concentration (0.1%). At glucose concentrations of 0.5-3% oxalic, succinic, fumaric, malic, citric and gluconic acids were excreted. Oxalic acid concentration increased with time and with an increase of sugar concentrations. The amount of succinic, fumaric, malic, citric and gluconic acids were the highest on 21st day of growth and increased with the increase of carbohydrates concentration. Dependence of acid concentration on the contents of carbohydrates in the medium is typical for fungi (Sazanova et al. 2016). In the culture of *B. subtilis*, the amount of oxalic acid was very low, much less than in the co-culture with fungus. The influence of oxalic acid production on oxalate crystallization by fungi was shown in our previous studies (Sturm et al. 2015; Sazanova et al. 2016). Oxalic acid concentrations in culture of *B. subtilis* and in the co-culture of *B. subtilis* + *A. niger* are presented in Table 2.4.

One of the key factors of the influence of fungus on the metabolism of bacteria in our opinion is pH of the medium. The optimal pH value for bacterial EPS production of genus *Bacillus* lies between 7 and 8 (Kawai et al. 1992). The same pH values were observed in our experiment in liquid monoculture of *B. subtilis*. In liquid co-culture of *B. subtilis* + *A. niger* the pH values decrease with the increase of the amounts of excreted oxalic acid and by the 30th day of the experiment becomes lower than 7 and when the glucose amount was 30 g/L the pH value reached 3. So according to our experiments an alkaline medium favors the EPS excretion and oxalic acid can be excreted both in an alkaline and in an acidic medium.

Thus, metabolic products formed by fungi and bacteria are essential for the processes of modern mineral formation. The ratio between the concentrations of EPS and oxalic acid in the medium is a key factor for the predominant crystallization of calcite or oxalates. A detailed discussion on mineral formation under the action of fungal and bacterial communities is given in the section 4.

A review of the original and published data showed that the metabolism of *A. niger* (primarily the release of organic acids) is a labile process depending on the fungal strain and many environmental factors. The chemical properties of the medium can both contribute to the activation of the metabolism of *A. niger*, including the increase of the excretion of organic acids, and the inhibition of this process, which should significantly affect biogenic crystallization. Among the components of the medium that differently affect the development and the metabolism of *A. niger*, we can, first of all, highlight the concentration of sugar and the form of nitrogen, as well as the presence of calcium carbonate and heavy metals in the medium. The metabolic activity of *A. niger* + *B. subtilis* association is vastly different from the activity of the separate monocultures. The interaction of the fungus and bacteria depends on the cultivation conditions (liquid medium or moist chamber, glucose concentration, pH of the medium).

3. Morphogenetic Patterns of the Formation of Metal Oxalates under the Influence of *A. niger* on Different Mineral Substrates

Oxalic acid salts (oxalates) found on the surface of rocks and minerals in biofilms with the dominance of lichens are characterized by the general formula $Me^{2+}[C_2O_4]nH_2O$ (Me^{2+} - Ca, Cu, Mn, Mg, Fe, Zn, Pb) (Table 3.1). Oxalates are localized in lichen thalli on the contact boundary with the rock, as well as in the core layer of lichen thalli. Metal ions involved in their formation enter the biofilm from the underlying stone substrate. Oxalate ions appear in biofilms as metabolic products of lichens and microscopic fungi among which are active producers of organic acids (for example species of the genera *Aspergillus*, *Penicillium*, *Trichoderma*, etc.). The results of model experiments on the formation of oxalates in lithobiotic systems under the influence of fungus *A. niger* are discussed below.

3.1. Formation of Calcium Oxalate Hydrates

Calcium oxalates (monohydrate whewellite and dihydrate weddellite) are the most common biofilm minerals. They were found in the presence of lichens and microscopic fungi on the surface of different calcium-containing rocks: carbonates (marble, limestone, dolomite), silicates (granite, serpentinite, gabbro, basalt, diabase, sandstone) and apatite-nepheline ore (Table. 3.1). In model experiments involving the fungus *A. niger*, calcium oxalates were obtained by us in the Czapek-Dox liquid nutrient medium and in a moist chamber (look Chapter 1) on the surface of carbonate rocks (calcite marble, dolomite), as well as on apatite-containing substrates (natural and synthetic fluorine and hydroxyapatite, apatite ore, dental tissue of fossil animals) (Sturm et al. 2015; Rusakov et al. 2016; Frank-Kamenetskaya et al. 2019; Zelenskaya et al. 2020).

3.1.1. Oxalate Formation on the Surface of Carbonate Rocks

Experiments with homogeneous calcite marble in the Czapek-Dox liquid medium showed that the concentration of oxalic acid increases with time and leads to significant dissolution of marble after 4 days of treatment (Sturm et al. 2015). Fungal hyphae are fixed on the marble surface (Figure 3.1a). Marble particles within the fungal mycelium in solution appear strongly etched (Figure 3.1b).

Calcium oxalate crystallization starts between day 9 and day 13 (at pH = 5.0-5.5) with the formation of almost ideal dipyramidal and dipyramidal-prismatic (with dominant {101} pyramidal faces) tetragonal weddellite crystals. Their size varies from 2 μm to 23 μm (Figure 3.2a). The prismatic faces {100} are comparatively small (edge/length ratio [100]/[001] ≈ 10). Parallel-grown (or so-called multiple-headed) weddelite crystals are also observed (Figure 3.2b) in the mycelium in solution which gives evidence that already at this step of reaction the supersaturation of the solution was high enough.

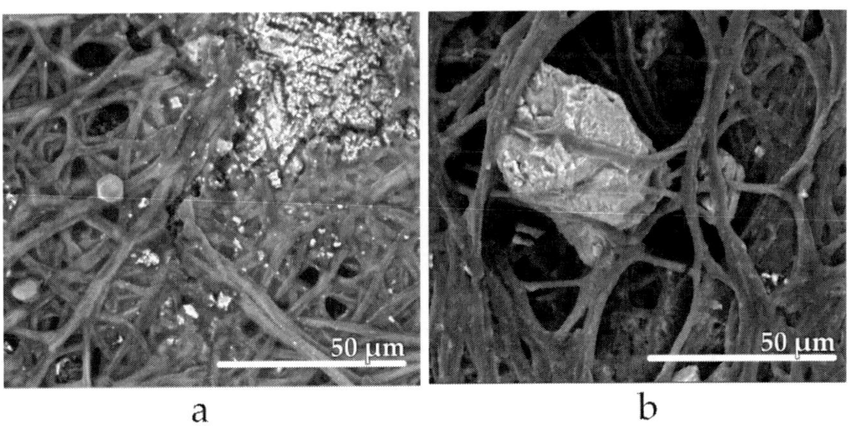

Figure 3.1. SEM images of *A. niger* mycelium formed in liquid medium after 4 days of experiment: a – on the surface of marble, b – in the solution.

Table 3.1. Oxalate minerals $Me^{2+}[C_2O_4]nH_2O$ in lihobiotic systems

Name	Formula		Sp.gr. sym.	Underlying rocks and minerals
	Me^{2+}	n		
Whewellite	Ca	1	*P*2$_1$/*c* (Izatulina et al. 2018)	(Ca, Mg) - carbonate rocks (Gehrmann and Krumbein 1994; Ascaso et al. 1982; Marques et al. 2016; Bungartz et al. 2004; Ríos de los et al. 2009; Syers et al. 1967; Souza-Egipsy et al. 2002; Rusakov et al. 2010); Ca-containing minerals of silicate rocks (Jones et al. 1980; Adamo et al. 1993; Wilson et al. 1981; Prieto et al. 1997; Bjelland et al. 2002); Gypsum is a weathering product of minerals of basaltic lavas and volcanic rocks (Chisholm et al. 1987; Purvis 1984); Apatite-nepheline ore (Frank-Kamenetskaya et al. 2019).
Weddellite	Ca	2.5–x, x≤0.5	*I*4/*m* (Mills and Christy 2016; Izatulina et al. 2018)	
Moolooite	Cu	0.5-1	*Pnnm* (O'Connor et al. 2019)	Copper and iron sulfides and their weathering products in basaltic lavas and more siliceous volcanic rocks (Purvis 1984; Chisholm et al. 1987; Purvis et al. 2008).
Lindbergite*	Mn	2	*C*2/*c* (Atencio et al. 2004; Baran 2014; Giester 1997)	Manganese ore, containing manganese oxides (Wilson and Jones 1984).
Glushinskite**	Mg	2		Serpentinite (Wilson et al. 1981, 1980).
Zinc oxalate*	Zn	2		In the vicinity of (Zn,Pb)S smelter (Sarret et al. 1998).
Lead oxalate*	Pb	0	*P-1* (Baran 2014)	

* Authors give chemical names.

** First discovered and described in lichens.

a b

Figure 3.2. SEM images of tetragonal weddelite crystals formed in liquid medium under the influence of *A. niger* after 13 days of experiment: a – on the surface of marble, b – in the solution mycelium.

Figure 3.3. SEM images of weddelite crystals and their intergrowths formed in liquid medium under the influence of *A. niger* after 16 days of experiment: a - the overview of the marble surface; b - dipyramidal-prismatic crystals with different sizes of the prismatic faces; c – «quadruplet» crystal intergrowth and skeletal crystal (inset); d - small dipyramidal crystals of subsequent generations on the surface of a large dipyramidal-prismatic crystal.

The remaining steps of calcium oxalate crystal formation (Figure 3.3, 3.4) took place at constant pH = 5. The prismatic faces of weddellite crystals became more pronounced (edge length ratio [100]/[001]) can reached 2) (Figure 3.3b), numerous intergrowths, skeletal forms (Figure 3.3c) and small dipyramidal crystals of subsequent generations appeared (Figure 3.3d). Calcium oxalate monohydrate whewellite appeared in the medium, as lamellar crystals, stack like intergrows and spherulite resembling aggregates (Figure 3.4) between 13 and 16 days of the experiment.

The appearance of whewellite does not lead to a decrease of weddellite amount (Figure 3.5a) this favors for a fast path of oxalate crystal formation and growth at this stage along with weddellite to whewellite transformation.

Under oligotrophic conditions of moist chamber, oxalate crystallization on the surface of marble proceeds much more slowly (Rusakov et al. 2016). Small weddelite crystals with prismatic faces developed much less than dipyramidal ones (edge/length ratio [100]/[001] ≈ 5) forms on homogeneous calcite marble only after one month of incubation (Figure 3.6a). After 4 months of incubation on marble surface the bigger weddelite crystals with more developed prismatic faces than those observed in liquid medium with the dissolution traces were observed (Figure 3.6b). There were no signs of whewellite in moist chamber.

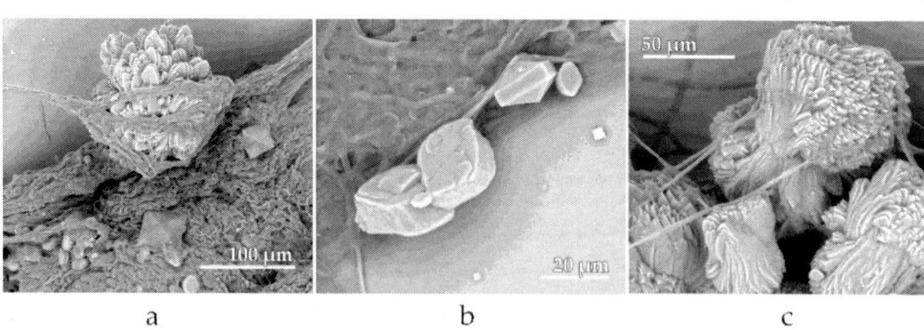

Figure 3.4. SEM images of whewellite crystals and their aggregates formed on hyphae of *A. niger* in solution mycelium on 16 day of reaction: a – the overview of the marble surface; b – lamellar crystals and their stack-like aggregates; c – spherulite-like intergrowth aggregates.

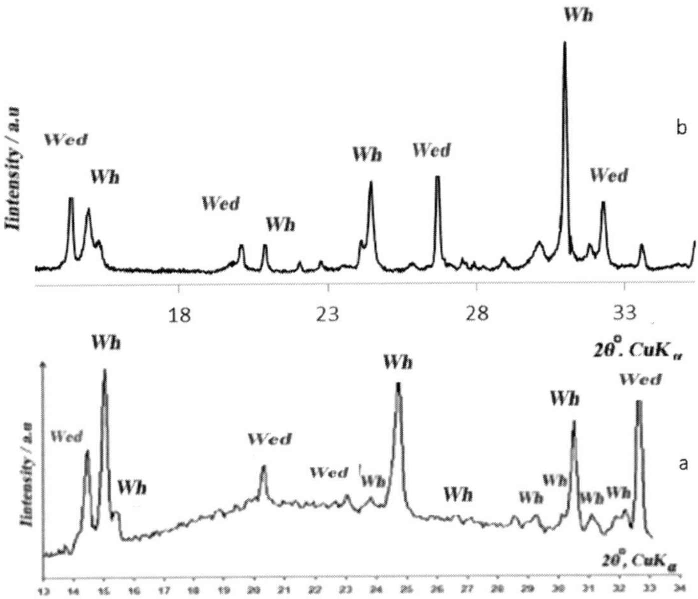

Figure 3.5. XRD pattern of calcium oxalates obtained in liquid medium on carbonate substrates by *A. niger* treatment: a – on homogeneous calcite marble after 16 days of reaction; b – on dolomite after 15 days of reaction. Key: *Wh* = whewellite, *Wed* = weddellite. Peaks of substrate minerals are also present (dolomite, quartz).

a b

Figure 3.6. SEM images of dipyramidal-prismatic tetragonal weddelite crystals formed on marble surface by *A. niger* treatment in moist chamber: a – after 1 month of experiment; b – after 4 months of experiment.

Figure 3.7. SEM images of weddelite crystals synthesized in liquid medium by *A. niger* treatment (a, b) and found in biofilms on the surface of marble monuments in Tauric Chersones, Crimea (c, d).

Morphology and size of the tetragonal weddellite crystals formed in a Czapek-Dox liquid medium on marble surface (Figure 3.7a, b) are closely related to crystals observed by us (Rusakov et al. 2010) on the surface of damaged Proconesos marble of monuments in Tauric Chersonesos, Crimea (Figure 3.7c, d). However, multiple-headed and skeletal weddellite crystals have not been observed in natural biofilms on the surface of carbonate rocks. This difference is probably due to the fact that the ionic supersaturation of the solutions and the crystal growth speed under the natural conditions are not as high as in our *in vitro* experiments.

Cultivation of the microscopic fungus *A. niger* in a liquid Czapek-Dox medium on the surface of dolomite led to the crystallization of calcium oxalates on the 7th day of the experiment. Weddellite is represented by dipyramidal-prismatic crystals and their intergrowths (40-50 µm) (Figure 3.8 a, b). The prismatic face is much more developed than the dipyramid face (edge/length ratio [100]/[001] ≈ 0.5). Whewellite forms lamellar small crystals ~10 µm in size (Figure 3.8b).

In the moist chamber, mass crystallization of calcium oxalates, mainly dipyramidal prismatic weddellite crystals (edge/length ratio [100]/[001] ≈ 1) was observed on dolomite surface after 30 days (Figure 3.8c). The formation of magnesium oxalates was not recorded in any of the experiments, which suggests the occurrence of magnesium ions as an isomorphic impurity in calcium oxalates, which is confirmed via EDX (Figure 3.9a).

Figure 3.8. SEM images of oxalates crystals formed under the influence of *A. niger* on dolomite surface: a – after 15 days of fungal treatment in liquid medium; b – after 21 days of fungal treatment in liquid medium; c – after one month of fungal treatment in moist chamber

3.1.2. Oxalate Formation on the Surface of Apatite-Containing Substrates

Cultivation of the microscopic *A. niger* fungus in Czapek-Dox liquid medium on the surface on apatite substrates (hydroxyapatite, fluorapatite, fossilized bones of animals) led to the crystallization of calcium oxalates, mainly monohydrate whewellite (Figure 3.10 – 3.14a), on the second day of the experiment at pH ~ 5 (Zelenskaya et al. 2020). Later the pH value reduced to 4. In syntheses with natural fluorapatite crystal - after 3 days, in syntheses with fluorapatite ore – after 5 days, in syntheses with fossilized dental tissue – after 6 days, and in syntheses with hydroxyl and fluorapatite powders – after 10 days.

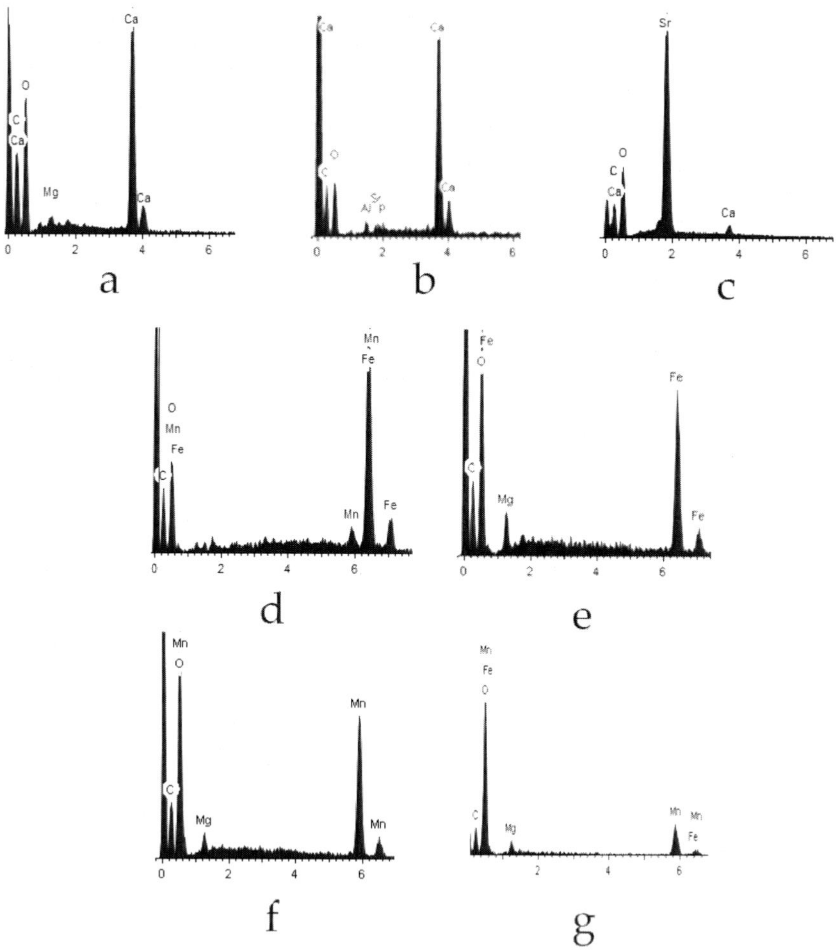

Figure 3.9. Impurities in oxalates obtained in a liquid Czapek-Dox medium under the influence of *A. niger* on various mineral substrates (EDX-data): a – magnesium in weddellite on dolomite, b – strontium in whewellite on apatite ore, c – calcium in Sr-weddellite on strontium carbonate powder, d – manganese in humboldtine on siderite, e – magnesium in humboldtine on biotite in rapakivi granite, f – magnesium in lindbergite on todorokite, g – magnesium and iron in lindbergite on kutnohorite.

The crystallization started earlier on the surface of the more soluble underlying substrate. The highest dissolution rate in organic acids secreted by the fungus among all substrates was revealed for synthetic hydroxyapatite and fluorapatite powders (higher for hydroxyapatite than for fluorapatite) (Driessens 1973; Bohner 2000; Wang et al. 2011). Dense fluorapatite crystal and almost monomineral fluorapatite ore have a significantly smaller surface area and, accordingly, their dissolution rate is lower than that of fluorapatite powder (judging by the intensity of oxalate crystallization). The solubility of fossil mammoth, horse and southern elephant teeth dentine, in organic acids secreted by fungus, is apparently less than that of apatite powders, but more than of fluorapatite ore and apatite single crystal, which can be well explained by the substantial porosity of dentine. The fact that the solubility of the dental tissue of a horse is greater than that of a mammoth and an elephant can be explained by the practical absence of the protein component in the fossilized tissue, as well as by the insignificant fluorine content in biogenic apatite.

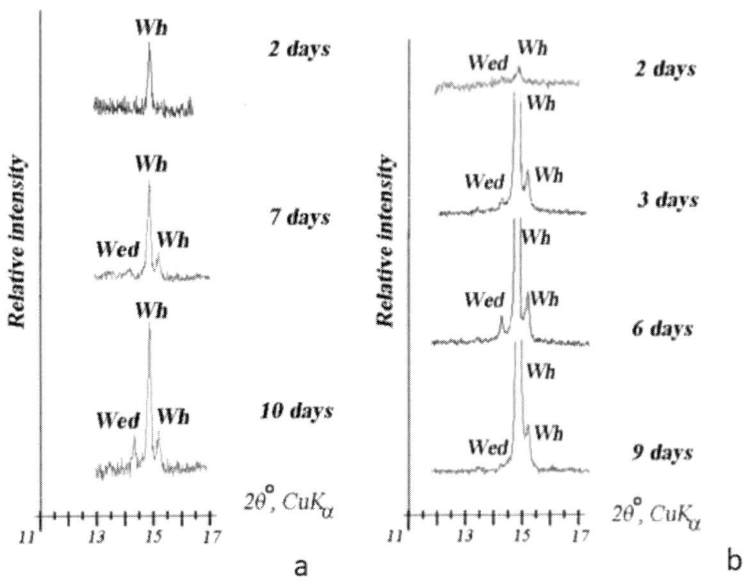

Figure 3.10. X-ray pattern fragments of oxalate crystallization products at various stages of the experiment: a – on the surface of hydroxyapatite powder, b – on the surface of elephant tooth.

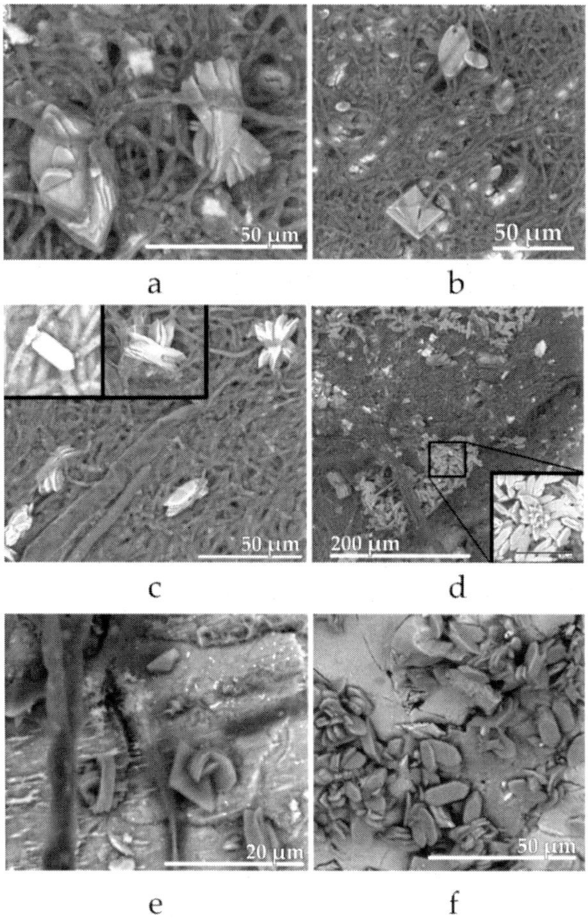

Figure 3.11. SEM-images of oxalate crystals formed by the action of *A. niger* on the surface of apatites in the initial stages of the experiment: a – lamellar split pseudohexogonal whewellite crystals and fan-shaped intergrowths of dovetail twins on powders of synthetic hydroxyapatite (on 2 day of the experiment); b – dipyramidal weddellite crystal among numerous small lamellar pseudohexogonal whewellite crystals on powders of synthetic hydroxyapatite (on 2 day of the experiment); c – individual crystals of whewellite, their intergrowths and dovetail-type twins on powders of synthetic fluorapatite (on 2 day of the experiment); d –pseudohexogonal whewellite crystalline aggregates of two generations on the powders of synthetic fluorapatite (on 3 day of the experiment); e – whewellite intergrowths, twins and separate dipyramidal crystals of weddellite on fluorapatite single crystal (on 2 day of the experiment); f – separate lamellar whewellite crystals among numerous aggregations of lamellar whewellite crystals and dovetail-type twins on fluorapatite single crystal (on 3 day of the experiment).

Lamellar pseudohexogonal whewellite crystals with straight or curved edges were the first crystalline phase to form on all apatite substrates (Figure 3.11, 3.13a). Probably local solution supersaturations lead to their intensive splitting and the formation of dovetail-type twins (Figure 3.11a,c,e,f) and various intergrowths: in the form of stacks of lamellar crystals (Figure 3.11a,c,d; 3.12b; 3.13a-inset), rose-like (Figure 3.11d-inset; 3.13a-inset), dumbbell-like (Figure 3.12a) and spherulite like aggregates (Fig3.12a; 3.13c,d; 3.14a). Lamellar crystals of new generations of whewellite formed at all stages of crystallization (Figure 3.11d; 3.12b). In addition, on the surface of hydroxy- and fluorapatite, lamellar crystals of whewellite and then their twins formed on various stages of crystallization (Figure 3.11f; 3.12c-inset). Mainly dipyramidal (3.11b,e; 3.12a,b; 3.13) and less commonly dipyramidal-prismatic (Figure 3.12c; 3.13d) weddellite crystals formed along with whewellite on the surface of all substrates but in smaller amounts.

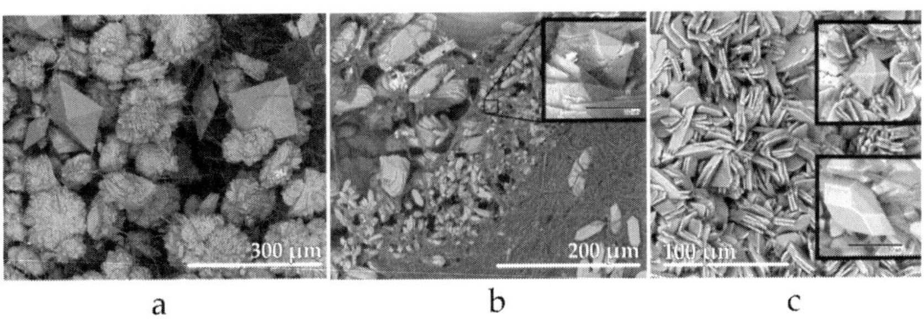

Figure 3.12. SEM-images of oxalate crystals formed by the action of *A. niger* on the surface of apatites at the final stage of the experiments: a – numerous spherulites and dumbbell-like whewellite aggregates and also large separate dipyramidal crystals of weddellite on powders of synthetic hydroxyapatite (after 10 days of experiment); b – whewellite crystals of various generations, their intergrowths and aggregates and separate weddellite crystals among them (inset) on powders of synthetic fluorapatite (after 10 days of the experiment); c – a carpet of whewellite dovetail-type twins on the fluorapatite single crystal, insets – dipyramidal-prismatic weddellite crystal and columnar whewellite crystal (after 9 days of the experiment).

Almost all crystals, twins, and intergrowths are characterized by curved edges on all hydroxy- and fluorapatite powders already on day 2 (Figure 3.11a,c,e). This indicates that the crystallization of oxalates began before day 2 and took place at high solution supersaturations, which led to an intense splitting of the crystals. On the single crystal of apatite, the amount of oxalates is much smaller, and their size is also smaller than on the powders (around twice as small) (Figure 3.11e). This indicates a less intensive crystallization process, which can be explained by a lower solubility of the substrate compared to the powders. At the same time, there are signs of both intense splitting and dissolution of crystals, which most likely indicates significant fluctuations in the supersaturation of the solution and possible changes in crystallization conditions, for example, a local pH decrease.

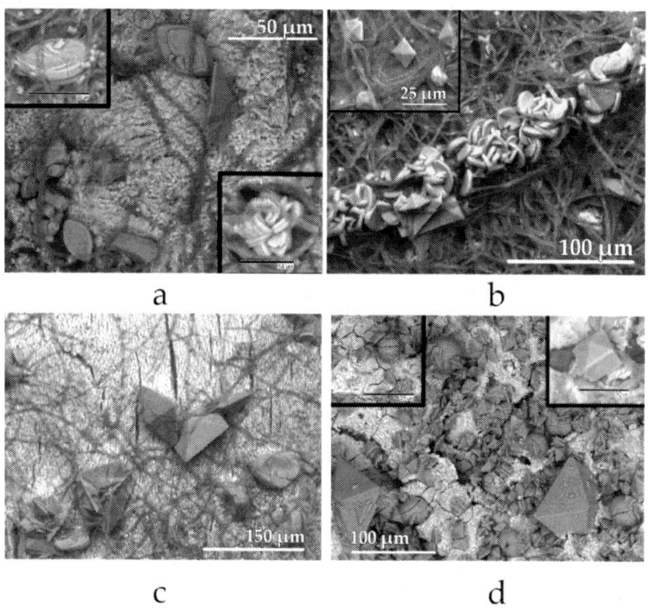

Figure 3.13. SEM-images of oxalate crystals formed by the action of *A. niger* on the surface of the mammoth tooth dentine at different stages of the experiment: a – lamellar whewellite crystals, their intergrowths (in the form «stacks» and «rosettes», see insets) and separate dipyramidal weddellite crystals of two generations (on 2 day of the experiment); b – numerous, associated with fungal hyphae, intergrowths of whewellite lamellar crystals as well as crystals of weddellite of various generations (on 3 day of the experiment); c – numerous spherulite-like aggregates of whewellites and large dipyramidal weddellite crystals (on 9 day of the experiment); d – spherulite-like box-shaped aggregates of whewellites and large dipyramidal crystals of weddellite; small dipyramidal-prismatic crystals of another weddellite generations – inset (on 13 day of the experiment).

Significant differences in the phase formation on the surface hydroxy- and fluorapatites under the action of fungus was observed as they exhibited different rates of weddellite formation. The formation of dipyramidal weddellite crystals was observed on hydroxyapatite powders on day 2 (Figure 3.11 b). On the 10th day, the size of individual weddellite crystals reach 180 μm (Figure 3.12a). On the surface of fluorapatite powders separate medium-sized weddellite crystals (~30 μm in size) were observed only on the 10th day of the experiment (Figure 3.12b). On the surface of fluorapatite crystal small separate crystals of weddellite were observed on days 2 and 9 (Figure 3.11e; 3.12c). In this case, on day 2, crystals have a dipyramidal habit, on day 9 — a dipyramidal-prismatic one. Weddellite did not form under the influence of *A. niger* in the synthesis on fluorapatite ore (Figure 3.14a).

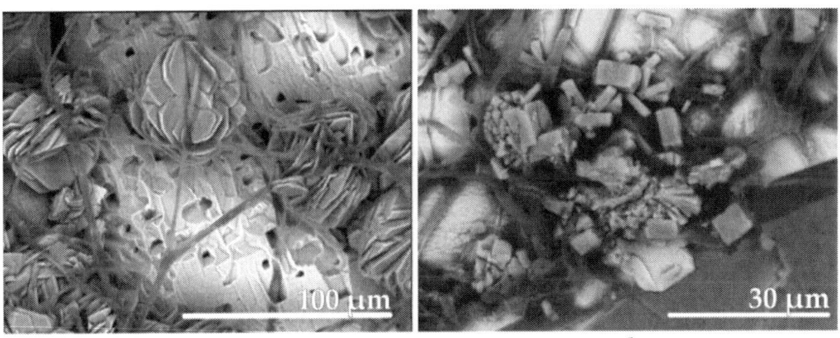

Figure 3.14. SEM images of calcium oxalates formed on fluorapatite ore: a - synthesized under the action of *A. niger* spherulite-like aggregates of whewellite crystals (on 5 day of the experiment); b – weddellite crystals with variously developed prism faces and fine-grained whewellite aggregates in biofilm with lichen *Lecanora polytropa* and microscopic fungi.

The peculiarities of calcium oxalate crystallization on the surface of mammoth tooth dentine (hydroxyapatite-protein composite) (Figure 3.13) were very close to those observed on hydroxyapatite powders (Figure 3.11ab; Figure 3.12a), which was probably due to a very porous structure of dentin. Various intergrowths of lamellar whewellite crystals exhibiting splitting and curved edges were observed already on day 2 (Figure 3.13a).

The dimensions of these crystals were similar to the sizes of the intergrowths formed on hydroxyapatite powders (Figure 3.11a). In addition, dipyramidal crystals of weddellite of two generations were observed. The formation of new generations of weddellite was also observed on the later stages of the experiment (Figure 3.13bcd). The largest weddellite crystals on days 9 and 13 of the experiment (Figure 3.13cd) were approximately two times smaller than on hydroxyapatite powders. Other distinctive features of crystallization on the mammoth dental tissue were the predominance of dipyramidal prismatic weddellite crystals on the 13th day of the experiment (Figure 3.13d). Spherical box-shaped aggregates (up to 80 μm) were seen on the surface of mammoth dentine on the 6, 9 and 13 days of the experiment (Figure 3.13c,d). We also observed the formation of such aggregates on the surface of hydroxyapatite powder on the 7th, and 10 days (Figure 3.12 a). The patterns of calcium oxalate crystallization on the surface of horse and elephant dentin are similar to each other and were close to those observed on the surface of mammoth tooth (Zelenskaya 2020, 596, 598), which is apparently due to the fact that the content of fluorine ions is small in apatite of these two fossils. Most quickly oxalates crystallized on horse tooth dentine, the IR spectrum of which did not show any traces of protein component: spherical box-shaped aggregates formed on the 3 day of the experiment and on the 9 day the largest reached 150-180 μm. The size of the largest dipyramidal weddelite crystals on the 9 day of the experiment was practically equal to the size of the largest weddellite crystals on hydroxyapatite powder. Crystallization on the surface of elephant tooth dentine was slower than on the mammoth tooth dentine. On the 2 day of the experiment, traces of substrate dissolution, separate lamellar crystals of whewellite with curved edges and dipyramidal weddelite crystals were clearly visible. On later stages, the size of whewellite crystals increased very slightly, their habit did not change, fully formed spherulite-like aggregates did not form (small aggregates were present in the form of rose-like aggregates). The size of the largest weddellite crystals on the 9 day of the experiment was slightly smaller than on the mammoth tooth (Zelenskaya et al. 2020, 598).

Thus, morphogenetic pattern of calcium oxalate crystallization under the action of the *A. niger* on the surface of all apatite substrates and on the surface of calcite marble and dolomite differ primarily by the predominance of whewellite over weddellite in the first one at all stages of the experiment, this is also supported by the published data (Adeyemi and Gadd 2005, 275; Pinzari et al. 2013, 1054, 1055). In experiments on apatite substrates, the value of the whewellite to weddellite ratio in the synthesis products increased as the solubility of the underlying substrate decreased, i.e., as the calcium content in the crystallization medium decreased. A much lower dissolution rate of a single crystal of fluorapatite and fluorapatite ore contributes to a smaller amount of calcium cations in the solution, which determines the formation of a very small number of weddellite crystals (on fluorapatite crystal, Figure 3.11e, 3.12c) or their complete absence (on fluorapatite ore, Figure 3.14a).

Phase composition of calcium oxalate crystals, synthesized on the surface of apatite ore under the action of *A. niger* (Figure 3.14a), differed significantly from the phase composition of calcium oxalates found in biofilm with lichen *Lecanora polytropa* and microscopic fungi (from the dumps of Khibiny Peralkaline Massif) (Figure 3.14b). In the experiment on the surface apatite ore forms only calcium oxalate monohydrate whewellite, containing admixture of strontium (Figure 3.9b), coming from underlying apatite rock (Frank-Kamenetskaya et al. 2019, 9). Whewellite crystals are characterized by multiple splitting and the formation of spherulite-like aggregates. Calcium oxalates in biofilm are represented by weddellite and whewellite. Weddellite is characterized by well faceted dipyramidal-prismatic and dipyramidal crystals, whewellite - by fine-grained aggregates. Identified differences indicate strong influence of the metabolic activity of the lithobiotic community organisms, which differs significantly from the activity of its individuals (look Chapter 2) on biofilm oxalate crystallization.

3.2. Formation of Strontium Oxalate Hydrates

Strontium oxalates were not yet found in nature. In model experiments involving the fungus *A. niger*, strontium oxalates were obtained by Sayer and Gadd (1997, 660) using Malt extract agar medium and by us on the Czapek-Dox liquid nutrient medium. In our experiments on a commercial strontium carbonate powder, crystallization of strontium oxalates began at pH = 4.5 after 1-2 days of the experiment (Figure 3.15a).

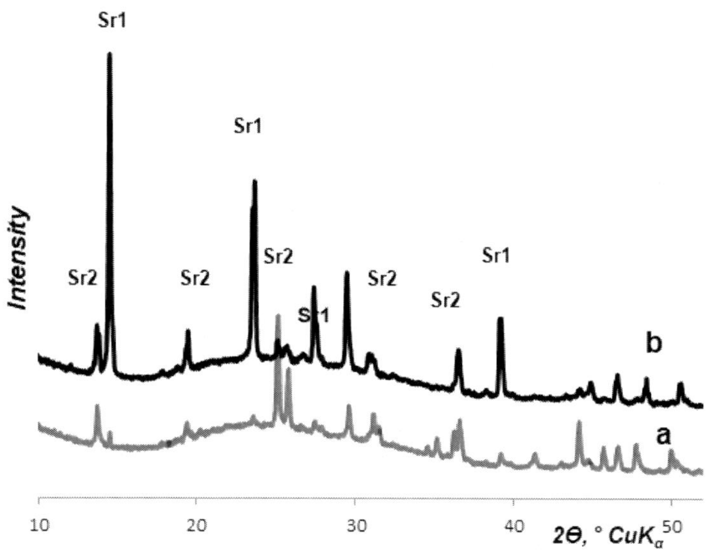

Figure 3.15. Fragments of X-ray diffraction patterns of strontium oxalates obtained in a liquid Czapek-Dox medium under the influence of *A. niger* on strontium carbonate powder at different stages of the experiment: a – 2 days, b – 6 days. Key: Sr1- strontium oxalate monohydrate, Sr2- strontium oxalate dihydrate

First, strontium oxalate dihydrate, which is a structural analogue of tetragonal calcium oxalate dihydrate weddellite, formed (Sterling 1965; Christensen and Hazell 1998; Baran 2014; Rusakov et al. 2019). After 3 days of incubation, the crystallization products were represented by numerous dipyramidal crystals of Sr-weddellite (from 5 to 25 μm) (Figure 3.16a). Among them, there were individual dipyramidal-prismatic crystals (edge/length ratio [100]/[001] ≈ 8), as well as large intergrowths of plate crystals (from 25 to 50 μm) of triclinic strontium oxalate monohydrate

(Figure 3.16a-inset). Numerous traces of dissolution were visible on Sr-weddellite crystals. Subsequently, the amount of strontium oxalates increases (Figure 3.15 b). The sizes of strontium oxalates monohydrates and dihydrates crystals also increased. On the 6 day of the experiment, the size of the crystals of strontium oxalate dihydrate reaches 100 μm, and the sizes of the strontium oxalate monohydrate intergrowths reach 200 μm (Figure 3.16b). On large crystals of Sr-weddellite, there are small (from 5 to 20 μm) dipyramidal Sr-weddellite crystals of the following generations. EDX data indicate the presence of a small calcium impurity in strontium oxalates (Figure 3.9c).

Figure 3.16. SEM images of strontium oxalates crystals formed under the action of *A. niger* on powders of commercially available strontium carbonate: a – 3 days of reaction; b – after 6 days of reaction.

In intensity crystallization of strontium oxalates on the surface of powder substrates is comparable to the crystallization of calcium oxalates on fluorine and hydroxyapatite powders and significantly exceeds oxalate crystallization observed on denser substrates (single crystal of natural fluorapatite, apatite ore etc.), which confirms the direct dependence of this process on the solubility of the underlying substrate. Strontium oxalate crystals obtained on an agar medium (MEA) with the addition of strontium nitrate powder under the action of *A. niger* fungus are characterized by a dipyramidal-prismatic habit common for dihydrous Sr-weddellite (Sayer and Gadd 1997). The ratio between the faces of the prism and the pyramid is close to the one obtained in our experiments after 3 days of incubation (Figure 3.16a).

As shown above, experiments conducted with the participation of the fungus *A. niger* allowed to detect the impurities of Sr^{2+} ions in calcium oxalates (Figure 3.9b), as well as impurities of Ca^{2+} ions in strontium oxalates via EDX spectroscopy (Figure 3.9c). Besides that, strontium impurities in weddellite and whewellite crystals in lichen thalli on Sr-bearing apatite rock were found (Frank-Kamenetskaya et al. 2019). This data suggests the possibility of the formation of $(Ca, Sr)[C_2O_4]\cdot nH_2O$ solid solutions (n = 1,2). The experiments carried out without the participation of fungi, allowed to prove the presence of two series of solid solutions: isomorphous $(Ca,Sr)[C_2O_4]\cdot(2.5\text{-}x)H_2O$ (sp. gr. I4/m) and isodimorphous $Ca[C_2O_4]\cdot H_2O$ (sp.gr. $P2_1/c$) – $Sr[C_2O_4]\cdot H_2O$ (sp.gr. $P\bar{1}$) (Rusakov et al. 2019).

3.3. Formation of Copper Oxalates Hydrates

Copper oxalate monohydrate moolooite $Cu(C_2O_4)\cdot H_2O$ was found in lichen thalli of *Lecanora polytropa, Lecidea lactea, Lecidea inops, Acarospora rugulosa* on copper and iron sulfides (chalcopyrite, pyrite) and weathering products (attackamite, brochantite) in basaltic lavas and more siliceous volcanic rocks (Table 3.1). In the lichen thallus of *Lecidea inops* on malachite, formed on the surface of chalcopyrite (Voronov Bor deposit, Karelia), moolooite is represented by small lamellar crystals 1-3 microns in size and their intergrowths (Figure 3.17). Crystal individuals can reach 5 microns.

In model experiments involving the fungus *A. niger*, moolooite was obtained on cuprite on Malt extract agar (Sayer et al. 1997, 33) and by us in Czapek-Dox agar medium (Figure 3.18). In addition, moolooite and other copper oxalates (fiemmeite $Cu_2(C_2O_4)(OH)_2\cdot 2H_2O$ and wheatleyite $Na_2Cu(C_2O_4)_2\cdot 2H_2O$) were obtained by us in a liquid Czapek-Dox medium on the surface of cuprite, natural and artificial malachite, and also on powders of synthetic cuprite and brochantite (Figure 3.19; 3.20).

Figure 3.17. SEM images of moolooite in the lichen thallus of *Lecidea inops* on malachite formed on the surface of chalcopyrite (Voronov Bor deposit, Karelia).

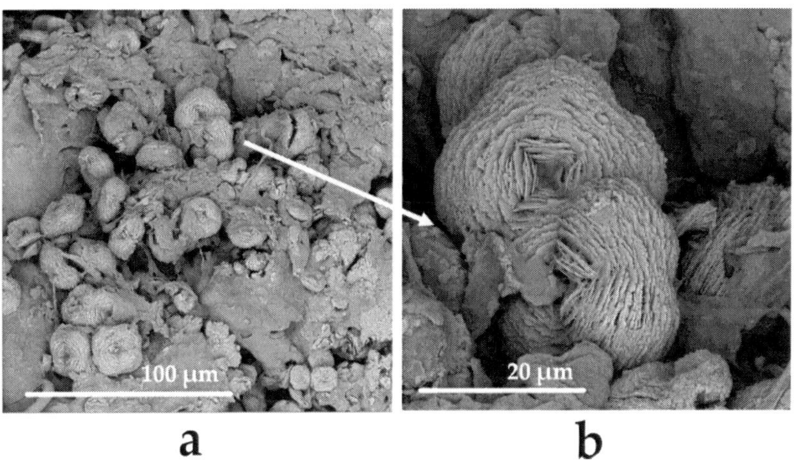

Figure 3.18. SEM images of moolooite spherulites formed on cuprite under the action of *A. niger* fungus in Czapek-Dox agar (after 5 days of the experiment).

Fiemmeite together with moolooite formed on the surface of natural malachite (Figure 3.19b). Previously, this mineral was never found in biofilms. It was first discovered and described as a new mineral (Demartin et al. 2018) in coalified woods at the base of the Val Gardena Sandstone (San Lugano, Val di Fiemme, Trento, Italy) which were permeated by mineralizing solutions containing Cu, U, As, Pb and Zn. The authors of this paper assume that oxalate anions have originated from diagenesis of the plant remains included in sandstones. Wheatleyite under the influence of *A. niger* formed together with moolooite on a synthetic brochantite powder (Figure 3.19c). It was found in the mine dumps of a Pb-Zn vein deposit (Wheatley mine, near Phoenixville, Pennsylvania, USA) (Rouse et al. 1986) and had not been found in biofilms.

Moolooite obtained under the influence of *A. niger* is represented on all substrates by "box-like" spherulite intergrowths which in turn are formed by perpendicular stacks of plate-like crystals the size of which, regardless of the underlying substrate, varies from 10 to 50 μm (Figure 3.18; 3.20).

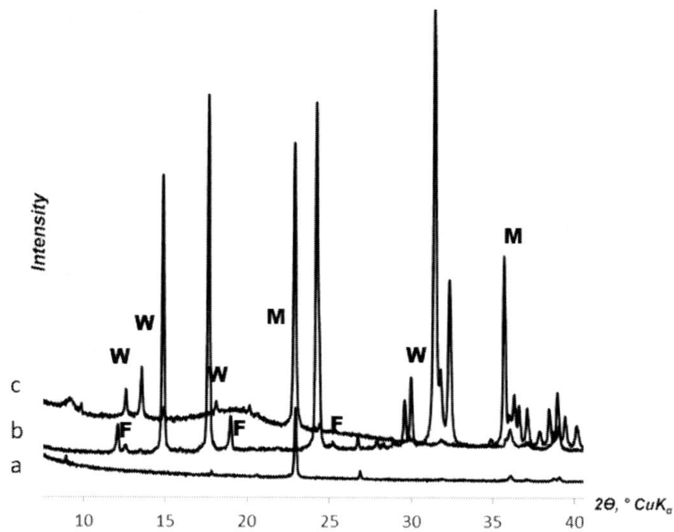

Figure 3.19. Examples of XRD patterns of copper oxalates obtained under the influence of *A. niger* in a liquid Czapek-Dox medium: a – on synthetic malachite, b – on malachite from the Urals, c – on synthetic brochantite powder. Diffraction maxima of malachite (a,b) and brochantite (c) are also present on the pattern. Key: W – wheatleyite; F – fiemmeite; M – moolooite.

In our experiments numerous intergrowths of moolooite crystals were observed on all Cu-bearing substrates already on the 5th day of fungal treatment at pH = 3.5-4 (Figure 3.20a,b). The size of individual lamellar crystallites varies from 5 to 10 μm, of intergrowths - from 10 to 30 μm. Then, during cultivation, the pH did not change (pH~4), the number of intergrowths was constantly growing, they uniformly covered the entire surface of the underlying substrate (Figure 3.20c, d). On top of large intergrowths (30–50 μm) and in the intervals between them, individual small lamellar crystals (5–10 μm) of subsequent generations (Figure 3.17) and their intergrowths (10–15 μm) similar to natural ones were observed.

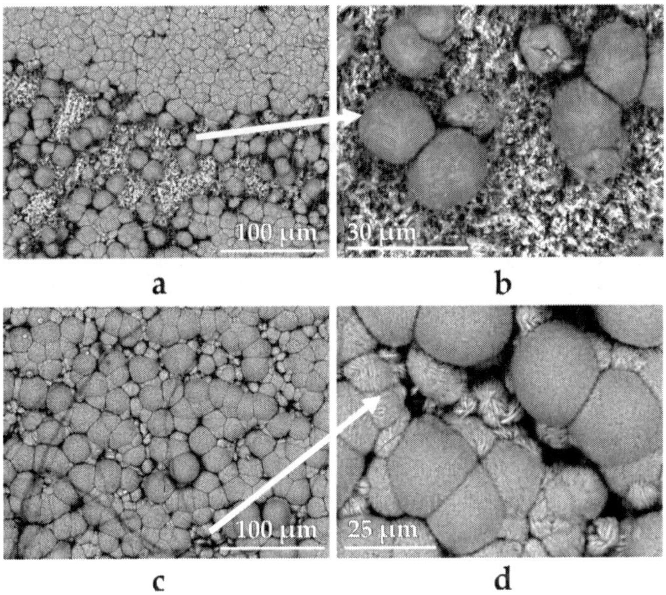

Figure 3.20. SEM images of moolooite spherulites in Czapek-Dox liquid medium, formed on the surface of natural malachite under the action of *A. niger*: a, b – after 5 days; c, d – after 12 days.

The appearance of numerous spherulite-like intergrowths of moolooite crystals in the syntheses under the influence of *A. niger* indicated higher crystallization rates due to high supersaturation of the solution, which lead to crystal splitting (Punin and Shtukenberg 2008).

3.4. Formation of Iron Oxalate Hydrates

Iron oxalate dihydrate humboldtine $Fe(C_2O_4)\cdot 2H_2O$ is one of the oldest known oxalate minerals (discovered in 1821) occurring on fracture surfaces in coal deposits and rarely in granitic pegmatites and hydrothermal mineral deposits.

Humboldtine has not yet been found in biofilms. Ascaso et al. (1982) reported about non-hydrated Fe (III) oxalate found at the boundary between dolomite rock and the lichen species *Calloplaca callopisma* and *Diploschistes ocellatus*. Unfortunately, the XRD presented in this work does not correspond to any of the known ferric oxalates, including non-hydrated Fe (III) oxalate (according ICDD PDF-2 Database (release 2016)). Several authors have disputed this finding before (Adamo and Violante 2000; Chen et al. 2000), because it is probable that iron (III) oxalate forms principally soluble oxalate complexes (Baran and Monje 2008). At the hypothesis level, it was suggested that it is impossible to find humboldtine in biofilms, since in biofilms a microbial oxidation of Fe (II) to Fe (III) occurs, followed by hydrolysis and precipitation of iron (III) oxides or oxo-hydroxides (Adamo and Violante 2000; Monje and Baran 2004). This assumption is supported by the findings of iron oxides in the thalli of lichens *Stereocaulon vesuvianum*, *Acarospora sinopica*, *Acarospora smaragdula* 'f. *subochracea*' on Fe-containing rocks (Adamo et al. 1997; Purvis et al. 2008). We obtained iron oxalate dihydrate humboldtine under the action of *A. niger* fungus in the Czapek-Dox liquid nutrient medium on siderite (iron carbonate), and also between biotite mica layers in rapakivi granite at pH = 3.5 (Figure 3.21a, 3.22).

After 5 days of cultivation, on the surface of siderite, numerous elongated (6–10 μm along the extension axis) monoclinic humboldtine crystals of prismatic habit were visible (Figure 3.22a). Subsequently, their number and size increased (Figure 3.22b). In addition to elongated crystals, there were short-prismatic crystals with poorly developed prism faces, which inclined to the axis of extension (on them strokes are clearly visible). There were crystals with traces of dissolution and splitting. There were also quadruplets (see the inset in Figure 3.22b).

On the surface of granite, crystallization was slower. After 30 days of cultivation, numerous small (from 5 to 12 μm) short-prismatic humboldtine crystals were visible (Figure 3.22 c). According to the EDX data in the humboldtine formed on the siderite surface there is an impurity of manganese (Figure 3.9d), and in the humboldtine formed between the layers of biotite on rapakivi granite – impurity of magnesium (Figure 3.9e).

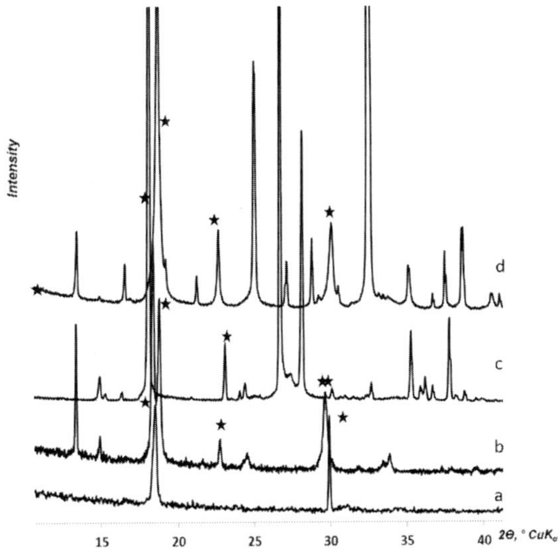

Figure 3.21. X-ray diffraction patterns of the humboldtine group oxalates obtained in the Czapek-Dox liquid medium under the action of *A. niger*: a – iron oxalate dihydrate humboldtine on siderite, b – manganese oxalate two-water lindbergite (diffraction maxima of falottaite and whewellite are also present on the pattern), c – zinc oxalate dihydrate (has no mineral analogues) on smithsonite (diffraction maxima of smithsonite are also present on the pattern). d– magnesium oxalate dihydrate glushinskite on magnesite. An asterisk indicates the main diffraction peaks of humboldtine group minerals.

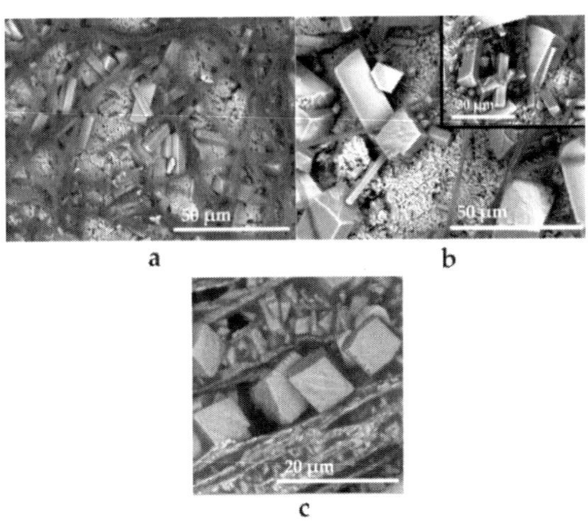

Figure 3.22. SEM images of humboldtine crystals and their intergrowths obtained under the action of *A. niger* in the Czapek-Dox liquid medium: a – on siderite, 5 days of cultivation, b – on siderite, 10 days of cultivation, c – between layers of biotite on rapakivi granite, 30 days of cultivation.

3.5. Formation of Magnesium Oxalate Hydrates

Magnesium oxalate was found in *Lecanora atra* lichen thalli on serpentinite (Insch, Kincardineshire and Island of Rhum, Inner Hebrides. Scotland) by Wilson et al. (1980, 1981) (Table 3.1) and was approved as the new glushinskite mineral in 1987. Named for Petr Ivanovich Glushinskii (1908-1990), coal geologist, Institute of Arctic and Antarctic Geology, St. Petersburg, Russia, who was the first to find this mineral in the core of well No. 944, the Chai-Tumus coal deposit on the river Lena (Yakutia, Russia).

Biofilm glushinskite represented with an abundance of crystals ranging from about 2 to 5 p μm in size (Wilson et al. 1980, 839, 1981 170). The majority of them show a distorted pyramidal (possibly pseudopyramidal) habit often with curved and striated faces. These crystals are dominated by magnesium but noticeable amounts of iron and traces of nickel and manganese can also be detected (Wilson et al. 1980, 840, 1981, 171).

In model experiments with the participation of fungus *A. niger*, magnesium oxalate dihydrate was obtained in a glucose solution on the surface of chrysotile, the serpentine group mineral (magnesium silicate with very low calcium content) (Wilson et al. 1981), in a liquid Czapek-Dox medium containing particles of forsterite (Figure 3.23) (Cai et al. 2013), as well as by us in the Czapek-Dox liquid medium on the surface of magnesite (Figure 3.21d, 3.24). Crystals of magnesium oxalate dihydrate obtained by Wilson et al. (1981, 173) on chrysotile under the influence of *A. niger* contain a significant admixture of iron, probably coming from a mineral substrate.

According to our data, individual short-prismatic almost isometric crystals of monoclinic glushinskite ~ 10-15 μm in size appeared on the surface of magnesite on the 18th day of the experiment at pH = 3 (Figure 3.24a). Almost all crystals show weakly developed prism faces (with clearly visible striation), which are inclined to the axis of extension in addition to the faces of the prism parallel to the axis of extension (similar to synthetic humboldtine crystals). Further, the pH did not change, the number of crystals increased, and by day 45 a continuous carpet of glushinskite crystals is formed, the sizes of which are 30-50 microns (Figure 3.24b). Short prismatic (same as in the syntheses on magnesite surface), glushinskite crystals were obtained under the influence of *A. niger* also by Wilson et al. (1981, 173) on the surface of chrysotile fibers and by Cai et al. (2013) on the surface of forsterite particles with a size of ~ 0.15-0.11 mm (Figure 3.23c, d). On the surface of larger particles of forsterite glushinskite (~ 0.42-0.25 mm) was observed with a habit of tubular prisms with a hollow interior chamber (Figure 3.23 a, b). Cai et al. (2013) on smaller forsterite particles (~ 0.15-0.11 mm in size) also obtained crystals of two-water magnesium oxalate of pyramidal-prismatic habitus (Figure 3.23d). The formation of such crystals, along with short-prismatic crystals of glushinskite, indicates the possibility for the formation of two forms of magnesium oxalate dihydrate under the action of *A. niger*: monoclinic (α) - a more stable form (Baran 2014) and orthorhombic (β) - morphologically similar to glushinskite crystals found in the thallus lichen (Wilson et al. 1980, 1981).

Figure 3.23. Glushinskite crystals obtained under the influence of fungus *A. niger* in liquid Czapek-Dox medium on forsterite particles after 10 days of incubation: a, b – size ~ 0,42-0,25 mm; c, d – size ~ 0.15-0,11 mm; a, c, d - SEM images, b - TEM image. Adapted from Cai et al. (2013) with kind permission of the author.

Figure 3.24. SEM images of glushinskite crystals obtained under the influence of *A. niger* in a liquid Czapek-Dox medium on magnesite: a – 18 days of cultivation, b – 45 days of cultivation.

3.6. Formation of Manganese Oxalates Hydrates

Manganese oxalate dihydrate was found in the thallus of the lichen *Pertusaria corallina* on manganese ore containing manganese oxides: lithiophorite and cryptomelane with fine intergrowths hollandite (Wilson and Jones 1984, 376) (Table 3.1). It was approved as a new lindbergite mineral only in 2003 after a detailed mineralogical description of crystals from Boca Rica mine, Galiléia, Minas Gerais, Brazil (Atencio et al. 2004). Manganese oxalate trihydrate, falottaite, is a new mineral found in the Falotta mine manganese deposit in the Swiss Alps by Graeser and Gabriel (2013, 2016) not found in biofilms yet.

Manganese oxalates under the action of fungus *A. niger* were obtained on malt extract agar medium containing rhodochrosite (Sayer et al. 1997, 33) and manganese oxides (Wie et al. 2012, 1747). In a liquid medium, manganese oxalates were obtained by Ferrier et al. (2019, 4) in AP1 medium with the addition of powders from manganese nodules as well as by us in the Czapek-Dox medium on the surface of Mn, Ca-bearing minerals: complex oxide todorokite $(Ca,Sr)_{0.3\text{-}0.5}(Mn^{4+},Mn^{3+},Mn^{4+},Mg)_6O_{12}\cdot 3\text{-}4.5H_2O$ (Figure 3.21b, 3.25) and carbonate kutnohorite $Ca(Mn,Mg,Fe)(CO_3)_2$ (Figure 3.26).

Figure 3.25. SEM images of magnesium oxalate intergrowths, obtained in liquid Czapek-Dox medium on the surface of todorokite under the influence of *A. niger* on the 18th day of experiment: a – intergrowth of lindbergite plate-like crystals growing through falottaite crystals, b – star shaped intergrowth of falottaite needle crystals.

Figure 3.26. SEM images of lindbergite intergrowths obtained on kutnohorite in liquid Czapek-Dox medium under the influence of *A. niger* on the 14th day of experiment: a – general view, b – spherulite-like intergrowth of plate-like crystals.

In all syntheses, manganese oxalate lindbergite was obtained as spherulite-like intergrowths of plate-like crystals (Figure 3.25a, 3.26), which in some cases were partially growing through much bigger falottaite crystals (Figure 3.25a). The star shaped intergrowths of falottaite needle crystals we obtained (Figure 3.25b) are similar to the intergrowths of manganese oxalate crystals described in Sayer et al. (1997). Numerous crystals of calcium oxalates (predominantly of whewellite) were also observed on the surface of todorokite and kutnohorite (on todorokite with strontium impurity).

According to EDX, manganese oxalates synthesized on todorokite contained small amounts of magnesium (Figure 3.9f), and on kutnohorite – impurities of magnesium and iron (Figure 3.9g) leaching from underlying mineral substrates. The isotypic structura of humboldtine group minerals (humboldtine, glushinskite and lindbergite) indicates that the impurities observed in their composition (Mg, Mn in humboldtine; Mn, Fe in glushinskite; Mg, Fe in lindbergite) replace Me^{2+} cations, resulted in formation of solid solutions with the general formula $(Fe,Mg,Mn)C_2O_4\cdot 2H_2O$.

3.7. The Formation of Zinc, Lead, Nickel and Lanthanum Oxalate Hydrates

Oxalates of zinc, lead, nickel and lanthanum have no mineral analogues. Zinc and lead oxalates were found in *Diploschister muscorum* lichen thallus in close vicinity of a (Zn,Pb)S smelter (Sarret et al. 1998) (Table 3.1). Nickel and lanthanum oxalates have not yet been detected in biofilms. Interest in the synthesis of these oxalates under the action of micromycetes is due to the potential application of microscopic fungi in bioremediation practices for various heavy metal contaminated media (see Chapter 5) and by industrial bioleaching and biorecovery of metals from ores.

Zinc oxalate was obtained in model experiments involving the *A. niger* fungus on a Malt extract agar containing zinc phosphate (Sayer and Gadd 1997, 657) and particles of zink sulfide mineral ores (Wei et al. 2013, 2177), as well as on PDA, containing zinc oxide (Sutjaritvorakul et al. 2013, 44). With the participation of *A. niger*, we obtained zinc oxalate dihydrate in the liquid Czapek-Dox nutrient medium on the surface of smithsonite (zinc carbonate) at pH = 3.5–4.0. (Figure 3.21c, 3.27). Obtained both in agar and in liquid media the two-water zinc oxalate, as well as the two-water manganese oxalate lindbergite (isotypic to it), forms spherolite-like intergrowths from plate-like crystals. In our experiments, their size during cultivation increased from 35-75 μm (18 days of the experiment) to 150-200 μm (30 days of the experiment).

Figure 3.27. SEM images of zinc oxalate dihydrate spherulite-like crystal intergrowths, obtained on smitsonite under the influence of *A. niger* in liquid Czapek-Dox medium on 18th day of experiment.

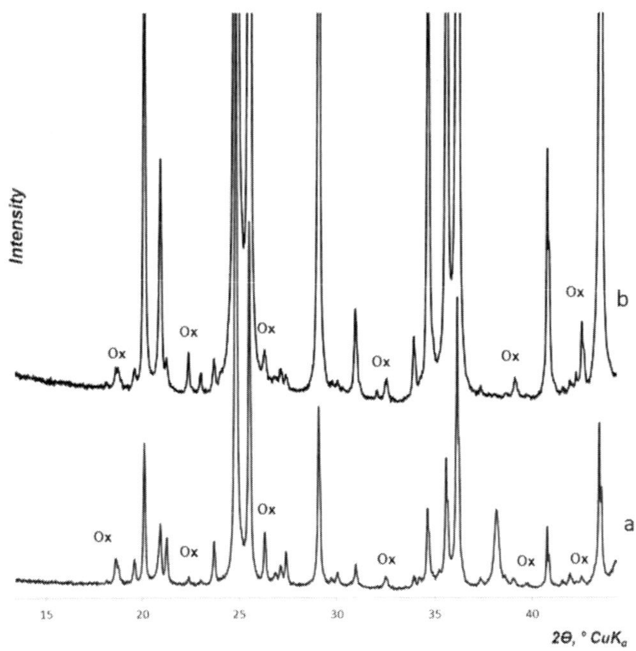

Figure 3.28. X-ray diffraction patterns of anhydrous lead oxalate (Ox) obtained by *A. niger* in Czapek-Dox liquid medium on cerrusite: a – after 7 days of cultivation, b – after 30 days of cultivation. Diffraction maxima of cerrusite are also present.

Lead oxalates under the action of *A. niger* were also obtained on agar media: Malt extract agar containing lead phosphate pyromorphite (Sayer et al. 1999, 692), Czapek-Dox containing lead arsenate mimetite (Ceci et al. 2015, 4958), PDA containing oxide and lead carbonate (Sutjaritvorakul et al. 2015, 59), lead sulfide (Adeyemi 2009, 480) and Pb foils (Wu et al. 2019, 3). Sayer et al. (1999) showed that under the action of *A. niger* lead oxalate dihydrate forms as intergrowths of needle crystals and anhydrous lead oxalate as intergrowths of plate-like crystals.

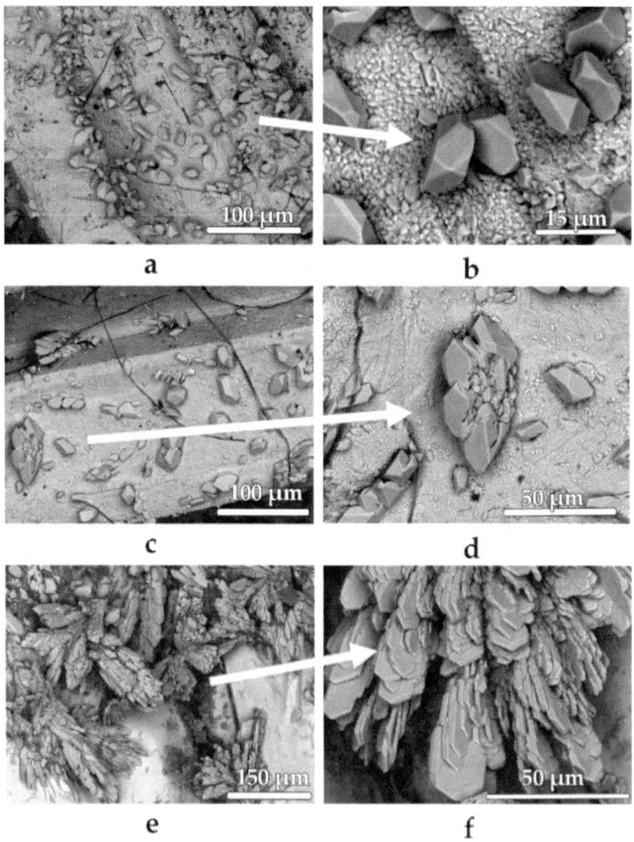

Figure 3.29. SEM images of anhydrous lead oxalate crystals and their intergrowths obtained on cerrusite under the influence of *A. niger* in Czapek-Dox liquid medium at various stages of the experiment: a, b – after 4 days, c, d – after 6 days, e, f – after 7 days.

Anhydrous lead oxalate was obtained by us under the action of fungus *A. niger* in liquid Czapek-Dox nutrient medium on the surface of cerussite (lead carbonate) at pH = 4 (Figure 3.28, 3.29).

On SEM images individual triclinic well-faceted crystals of anhydrous lead oxalate (15–20 μm) were visible (Figure 3.28a,b) as early as on days 2–4 of cultivation, further, the number of lead oxalate crystals increased and after 6 days of cultivation individual split crystals (Figure 3.28c,d) and later numerous spindle-shaped intergrowths of elongated multi-headed skeletal crystals (150-200 μm) were formed (Figure 3.28e,f).

Figure 3.30. SEM images of mycogenic La-oxalate crystal intergrowths obtained under the action of *A. niger* in solid Czapek-Dox medium with $LaCl_3$ (inset: optical microscopic image of the dendrite). Adapted from Kang et al. (2019) with kind permission of the author.

Nickel oxalate dihydrate was obtained by Magyarosy et al. (2002) in potato dextrose broth (PDB) containing $NiCl_2$ (a final concentration of 6.5 mM). Nickel oxalates in the form of small rectangular crystals were associated with the cell walls as well as localized inside the cell. Authors suggested that nickel ions penetrate the cell wall during cultivation of the fungus in the presence of nickel, and nickel oxalate crystals form as oxalate is excreted from the cell.

Lanthanum oxalate decahydrate was obtained under the action of *A. niger* on Czapek-Dox agar medium amended with $LaCl_3$ (Kang et al.

2019). Mycogenic La oxalates form dendrite intergrowths of tabular crystals and are characterized by a terraced surface (Figure 3.29).

3.8. Conclusion Remarks

The review of the results of model experiments on the formation of oxalates under the influence of the fungus *A. niger* on different mineral substrates showed that general morphogenetic laws of this process are determined by the chemistry of the crystallization medium and depend both on the fungal metabolic activity and on properties of underlying mineral substrate.

On the surface of mineral substrates inoculated by *A. niger* only oxalic acid salts (oxalates) are observed. The absence of salts of other organic acids secreted by the fungus (look Chapter 2) indicates their strong solubility in aggressive products of fungal metabolism. The presence of these aggressive compounds on the surface of the underlying substrates has a strong effect on oxalate crystallization, since they contribute to the dissolution of the substrate and strongly affect the pH of the crystallization medium. The process of oxalate formation is also significantly affected by the cultivation conditions of *A. niger* (look Chapter 1), since they significantly affect its metabolic activity (look Chapter 2). As shown by the example of calcium oxalates formed on the surface of calcite marble, oxalate crystallization under oligotrophic conditions of a moist chamber is limited by a lack of nutrient resources and is an order of magnitude slower than in a liquid medium.

The influence of the underlying substrate on the process of oxalate crystallization under the influence of *A. niger* is determined by its elemental and mineral composition, as well as structural features (porosity, density), i.e., from the parameters determining its solubility. The greater the solubility of the substrate, the faster crystallization occurs and the pH decreases more slowly. In addition, ions entering the crystallization medium from the underlying substrate can both accelerate and slow down the fungal metabolism (look Chapter 2), which also affects the crystallization rate. With time as the fungus culture ages the pH values in the crystallization medium begin to grow slowly. According to the experimental data discussed, crystallization of oxalates under the influence of *A. niger* occurs on various substrates at pH = 2.5–5.5 (Table 1.3).

In the case of calcium oxalates, a decrease in the solubility of the substrate, which leads to a decrease in the ratio of calcium ions to oxalate ions in the solution, changes the phase composition of the precipitate: the amount of calcium oxalate dihydrate weddellite decreases, and the monohydrate whewellite increases. Weddellite crystallization can be catalyzed by various organic acids (primarily citric acid) secreted by *A. niger* and inorganic impurities that enter the solution from the underlying substrate (e.g., Mg, Fe cations) (Kuzmina et al. 2019).

The components of the solution actively interact with crystallizing oxalates, either replacing Me^{2+} ions or adsorbing on the faces of growing crystals. The spectrum of ionic substitutions in metal oxalates is quite wide and depends on the elemental composition of the underlying stone substrate. To date, the patterns of strontium incorporation into calcium oxalates (monohydrate whewellite and dihydrate weddellite) are well studied (Rusakov et al. 2019). It is shown that the isotypic minerals of the humboldtine group group (humboldtine group, glushinskite, and lindbergite), obtained with the participation of the fungus *A. niger* on the surface of various substrates, are solid solutions with the general formula $(Fe,Mg,Mn)C_2O_4 \cdot 2H_2O$.

Morphological features of oxalate crystals formed under the influence of *A. niger* are well explained by selective adsorption of the organic and inorganic components of crystallization medium to the growing individual. For example, differently developed prism faces of weddellite formed on the surface of different Ca-bearing substrates are well explained by the preferential adsorption of admixture components (organic products of fungal metabolism, primarily citric acid, Mg^{2+}, Sr^{2+} ions, etc.) on these faces (Thomas et al. 2012; Sturm et al. 2015; Frank-Kamenetskaya et al. 2019; Kuzmina et al. 2019). Variations in the content and, probably, in the composition of impurities adsorbed on the prism faces lead to various degrees of inhibition of its growth and development. The inhibition process is very intensive on the dolomite bedrock, where weddellite crystals with significantly more developed prism faces than the dipyramid faces were formed. It is also likely that strong adsorption interaction of impurities with oxalate crystals during their growth lead to their intensive splitting and the formation of various regular and irregular intergrowths, as well as skeletal forms that are not found in biofilm on the surface of rocks. This distinction between the morphology of natural crystals and crystals, obtained *in vitro* experiments under the action of *A. niger*, is probably due to the fact that the ionic supersaturation of the solutions and the crystal growth speed under the natural conditions are not as high as in experiments. Oxalate crystals synthesized under *A. niger* action have close morphology to oxalates from natural biofilms in the following cases: at the early stages of crystallization (as weddellite formed on the surface of calcite marble), during the formation of small crystals of subsequent generations (as moolooite was formed on the surface of various copper substrates).

4. The Patterns of Crystallization on Marble by Action of *A. niger* + *B. subtilis* Association

Usually rock surfaces in nature are inhabited by the multi-species microbe communities (mainly fungi and bacteria) (Boonchan et al. 2000; Gilbert et al. 2002; Berdoulay and Salvado 2009). As discussed in chapter 2, the biochemical activity of microorganism communities differs significantly from monocultures of bacteria and fungi. Unfortunately, the patterns of microbial crystallization under the influence of several species of microorganisms are practically unexplored.

Below we give an overview of our study of mineral formation under the influence of oxalate-producing bacteria *B. subtilis* and *A. niger* + *B. subtilis* association in different experimental condition (Figure 4.1). (Sazanova et al. 2020).

4.1. Crystallization in Liquid Czapek-Dox Medium

Firstly we studied crystallization processes in monocultures of bacteria. The experiments in liquid Czapek-Dox medium showed that calcite crust formed on the marble surface under the influence *B. subtilis* (Figure 4.1a,c,e). An increase in glucose concentration (from 1 to 30 g/L) leads to a more intense formation of secondary calcite. At a low glucose concentration (1 g/L) on 30 day of cultivation a separate clusters of small calcite globules (less than 0.2 μm) on the marble surface are seen (Figure 4.1a). At the highest glucose concentration (30 g/l) the size of calcite globules reaches up 2-3 μm already on the 21 day of cultivation (Figure 4.1e). After 60 days of experiment glucose concentration 30 g/l a continuous calcite crust formed on the surface of marble.

There were no visible crystals of calcium oxalate in cultures of *B. subtilis* bacteria. A very small amount of calcium oxalate monohydrate whewellite was observed by means of X-ray powder diffraction in the precipitates received at different glucose concentrations in the medium (1-30 g/l) after 30 days of the experiment (Figure 4.2a). Apparently whewellite crystals were covered by a thick crust of secondary calcite so their morphology was obscured for the microscope.

In the liquid co-culture of *B. subtilis* + *A. niger*, the phase composition of the crystallization products changed depending on the concentration of glucose in the medium. The observations of marble surface during the log phase of microbe growth (21-30 days of cultivation) showed globules of secondary calcite (smaller than 0.2 μm) only at a low glucose concentration 1 g/L on 21 day of experiment by pH = 8 (Figure 4.1b). Along with calcite globules, calcium oxalate crystals were present on the surface of marble: tetragonal dipyramidal (almost isometrical) weddellite crystals and separate plate like monoclinic whewellite crystals. The size of weddellite crystals varied from 10 to 25 μm. Whewellite plate like crystals (5-40 μm) formed stack-like intergrowths.

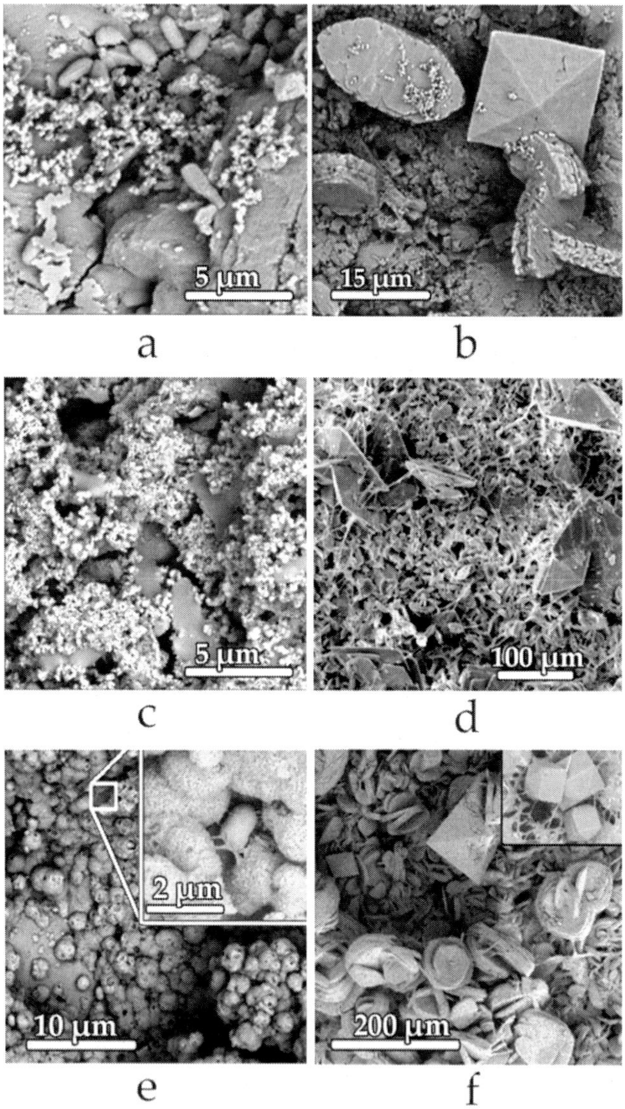

Figure 4.1. SEM images of marble surface after the treatment with bacteria *B. subtilis* (a, c, e) and *A. niger + B. subtilis* association (b, d, f) in liquid Czapek-Dox medium in log-phase of culture growth (a, c – 30 days of cultivation, b, d, e, f, - 21 days of cultivation) depending on the glucose concentration in the medium: a, b – 1 g/l; c, d – 10 g/l; e, f – 30 g/l; insets: e - separate bacterial cells between calcite globules, f – weddellite with a pronounced prism face.

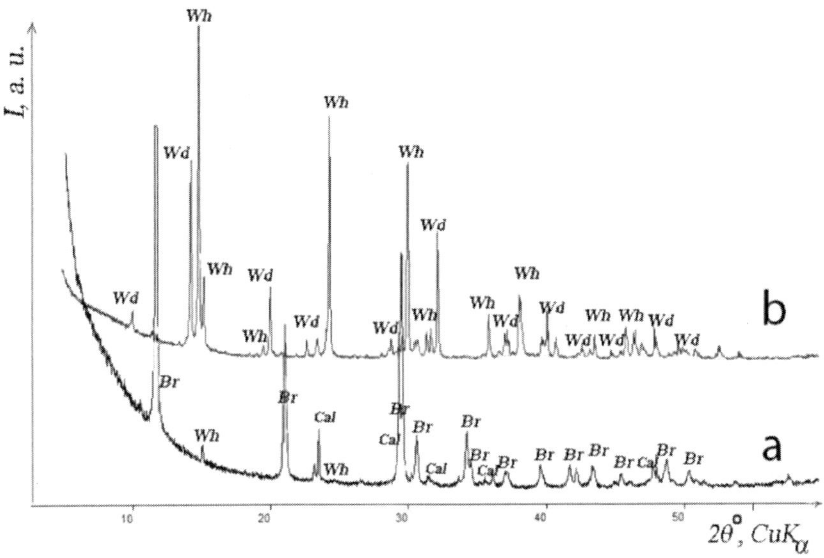

Figure 4.2. Powder X-ray diagram of crystallization products formed in liquid Czapek-Dox medium on marble under the influence of microorganisms: a – *B.subtillis* (30 days, 1 g/L glucose), b – *B.subtillis* + *A. niger* (21 days, 30 g/L glucose). Key: Wd – weddellite, Wh – whewellite, Cal – calcite, Br – brushite.

In the experiment with glucose concentration of 10-30 g/L (pH~5) secondary calcite was not observed (Figure 4.1d, f). At glucose concentration of 10 g/L massive dipyramidal weddellite crystals (up to 100 μm) stuck in the continuous carpet of whewellite crystals (up to 30 μm each) on 21 day of experiment were observed. At glucose concentration of 30 g/L, numerous plate like and spherulitic whewellite intergrowths (up to 80 μm) and weddellite crystals of approximately the same size were found. Some of the observed weddellite crystals had prismatic faces as well as dipyramidal ones.

In the monoculture of *B. subtilis* in liquid Czapek-Dox medium the amount of EPS increased with the increase of the amount of sugar in medium and was always higher than the amount of oxalic acid (Table 2.2) which resulted in pH increse and the precipitation of secondary calcite. The intensity of carbonate crystallization increased with the increase of sugar amount and consequently the amount of EPS in the medium (Figure 4.1).

The oxalic acid concentration in the medium before 30 days of the experiment was not enough for supersaturating the solution which would lead to the calcium oxalate crystallization. Only after 30 days of the experiment small amounts of calcium oxalate monohydrate (whewellite) were found in the crystallization products.

Thus, our data confirm that the process of calcium carbonate precipitation is usually controlled by calcium concentration as well as the concentration of dissolved inorganic carbon, pH value and the accessibility of nucleation sites (Muynck et al. 2010). Calcification process depends on EPS production by bacteria (Ercole et al. 2007; Dhami et al. 2013; Zhu and Dittrich 2016) which become the nuclei for calcite crystal crystallization in the solution (Cao et al. 2016).

The obtained data from the experiments with *B. subtilis* + *A. niger* in liquid Czapek-Dox medium demonstrated that metabolic activity of microscopic fungi can suppress the formation of bacterial EPS and prevent the formation of calcite crust. In liquid co-culture of *B. subtilis* + *A. niger* the ratio between EPS and oxalic acid concentrations in cultural broth is lower than in *B. subtilis* monoculture and decreases with the increase of the amount of sugar in the medium. The pH values decrease as well and become less than 7. When the amount of glucose was 1 g/L the EPS concentration was higher than the amount of oxalic acid by 200 times (at pH = 8) and calcite and calcium oxalates (weddellite and whewellite) crystallized. With the increase of glucose amount (≥ 10 g/L) oxalic acid concentration got substantially higher than EPS concentration (by ≥1000 times), pH value decreases to 5 and the conditions became favorable only for oxalate crystallization.

According to our results the variable phase composition of the crystallization products under influence of *B. subtilis* and *A. niger* + *B. subtilis* association is controlled by pH value which largely depend on the metabolites secreted by bacteria and fungi (EPS and oxalic acid). In turn, the acidity of the medium significantly affects the metabolism of microorganisms.

4.2. Crystallization in Moist Chamber

The observed patterns of crystallization in oligotrophic conditions of moist chamber significantly differ from the ones observed in liquid medium. In a monoculture of *B. subtilis* single dipyramidal crystals of tetragonal calcium oxalate dihydrate weddellite were observed on marble surface after 30 days of experiment (from 5 to 15 μm). After 3 months of cultivation the number of dipyramidal crystals did not increase, the size of the crystals ranged from 10 to 40 μm (Figure 4.3a). Bacterial cells were found both on the surface of the oxalate crystals and around them. There was no secondary calcite found on marble surface even after 90 days of incubation.

In the experiment with *B. subtilis* + *A. niger* co-culture in moist chamber the formation of weddellite crystals was observed between 21 and 90 days of the experiment. The ratio of prismatic to dipyramidal face size was different than in experiment with bacteria only: prismatic faces were considerably bigger than the dipyramidal ones (Figure 4.3b). Crystallite sizes ranged from 15 to 30 μm. The number of Ca-oxalate crystals formed on marble surface was almost the same for both experiments.

Figure 4.3. SEM images of weddellite crystals formed in moist chamber on marble surface after 90 days of incubation with microorganisms: a – *B.subtillis* (inset: dipyramidal weddellite crystals); b – *B.subtillis* + *A. niger* (inset: dipyramidal-prysmatic weddellite crystals).

Oligotrophic conditions in moist chamber are not favorable for EPS accumulation and the formation of secondary calcite. The products of biogenic crystallization on marble surface consisted of calcium oxalate dihydrate (weddellite). In the experiment with *B. subtilis* monoculture high concentrations of oxalate ions in moist chamber was induced only by the metabolism of *B. subtilis*. In the experiment with *B. subtilis* + *A. niger* co-culture the intensity of weddellite crystallization was practically the same as in the experiment with *B. subtilis* monoculture which means that the oxalate crystallization was inhibited due the lack of nutrient substances.

As shown above, bacteria, along with fungi, are able to produce oxalic acid, but in much smaller quantities. *Bacillus* spp. may also exhibit oxalotrophic activity, depending on the conditions (Tanner and Bornemann 2000; Schneider et al. 2012; Lee et al. 2014). However, oxalic acid consumption by bacteria is energetically disadvantageous and in presence of other carbon sources oxalic acid is not consumed (Bravo et al. 2013).

In the absence of sugar excess oxalate ions formed under the action of fungi, bacteria or plants get consumed by oxalotrophic bacteria, which leads to a local pH increase and, ultimately, to a change from oxalate to carbonate crystallization. Oxalate catabolism is carried out by bacteria of different taxonomic groups (Jenni et al. 1987, 1988; Sahin 2003). Oxalate catabolism plays an integral role in the oxalate-carbonate pathway, a complex biogeochemical process that occurs under the influence of biological and geological factors (Verrecchia et al. 2006).

In our experiments with *A. niger* + *B. subtilis* co-culture, we did not observe the change from oxalate to carbonate crystallization. However, as the calcium oxalates crystallized and the cultures of microorganisms aged, an increase in the initial pH value may have occurred (Sturm et al. 2015; Sazanova et al. 2016). This suggests that if the nutrient substances in the moist chamber still preserved the carbonate crystallization could take place after the oxalate crystallization. This assumption is confirmed by the metamorphose of secondary calcite over the splitted whewellite crystal observed by us in an earlier experiment (Figure 4.4) carried out in moist chamber with *B. subtilis* bacteria and *Penicillium* sp. fungus co-culture (Rusakov et al. 2016).

Figure 4.4. SEM images of metamorphose of secondary calcite over the splitted whewellite crystal, formed under the influence of *B. subtilis* bacteria and *Penicillium* sp. fungus co-culture in moist chamber: a – general view; b – spherulitic crystal intergrowth.

4.3. Conclusion Remarks

The present research showed that the crystallization patterns discussed above can be well explained by the characteristics of the metabolism of microorganisms in various conditions (look Chapter 2). This influence is different in bacterial monoculture and in bacterial-fungal associations.

Generalization of the experimental results on mineral formation under the influence of *A. niger* + *B. subtilis* association in different conditions demonstrate that crystallization under the action of bacterial-fungal associations differs significantly from that occurring under the action of individual microorganisms. In the *A. niger* + *B. subtilis* association, the formation of calcium oxalates or carbonates (carbonate-oxalate pathway) can be observed as a function of changing environmental conditions. In this case, the morphology of the crystals substantially depends on the composition of the metabolites produced by the fungus and bacteria.

The results obtained in experiments with individual species, including *A. niger*, should be used with caution when interpreting processes occurring in lithobiont systems with the participation of wide associations of microorganisms.

In addition, the possibility of suppressing bacterial carbonate crystallization by microscopic fungi must be taken into account when transferring conservation bacterial biotechnologies for protecting marble monuments from environmental influences from laboratory to natural conditions, taking measures to prevent contamination of the marble surface with ubiquitous microscopic fungi. As our experience (Frank-Kamenetskaya et al., 2019) shows, regular routine care associated with the chemical cleaning of the surface of marble monuments can significantly reduce the number of microorganisms (first of all, microscopic fungi) on rock surface. After such cleanin, it is possible to use "beneficial" bacteria to restore the surface of the monument, taking into account the specific parameters of the crystallization medium.

5. Perspectives of Application of *A. niger* in Applied Research

5.1. Heavy Metal Detoxication Mechanisms

Heavy metal pollution creates a significant threat to ecosystems. Unlike organic pollutants, heavy metals are not biodegradable and can be included in the food chain through bioaccumulation. Some metals, including Cu, Fe, Mn and Zn, are essential elements for organisms. However, when the concentration of these metals in the environment is very high, they become toxic (Siddiquee et al. 2015). It is important to study the resistance of fungi to these threads at the strain level to discover the physiological mechanisms of adaptation of fungi to stress and the possible use of fungi for cleaning the environment from toxic elements (bioremediation).

Bioremediation is the ability of living organisms to turn pollutants into a less toxic form. Toxic substances are transformed by living organisms as a result of reactions that occur as part of their metabolic processes. Using bioremediation is a safe, effective, and relatively inexpensive way to clean up contaminants (Malik et al. 2015; Siddiquee et al. 2015).

Fungi take part in the processes of metal solubilization and immobilization, which are key processes in biogeochemical cycles. The actions of fungi on heavy metal ions are aimed to detoxify them at the polluted sites and may occur through several mechanisms extracellularly or intracellularly (Siddiquee et al. 2015). Solubilization mechanisms, such as, binding of toxic elements to organic acids, other metabolites, and siderophore cellular proteins, can mobilize metals into forms accessible for cellular absorption. The immobilization of toxic elements can result from sorption on cell walls and exopolymers, as well as their intracellular and extracellular binding and deposition (Gadd 1996, 1997, 2001; Sayer and Gadd 1997; Gadd and Sayer 2000).

The metabolic products extracellularly excreted by fungi can increase the dissolution rate of the initial metal compounds, such as phosphates, and therefore lead to an increase in their mobility and bioavailability. Although fungal leaching can occur as a result of several processes, including siderophore formation (in the case of iron), leaching by most fungi appears to be mainly due to the production of organic acids, primarily citric acid (Sayer and Gadd 1997; Gadd 1999; Gadd and Sayer 2000; Gadd 2001).

Microbial solubilization may interfere with bioremediation. For example, lead phosphate – the mineral pyromorphite can be dissolved under the action of some fungi, for example *A. niger*, which will lead to the transition of lead into a mobile form (Sayer et al. 1999).

The immobilization of metals by fungi can occur both as a metabolically active process, and in dead biomass (Gadd 2001). The fungal cell wall interacts with metals, playing an important role as a protective layer and a barrier that controls the absorption of potentially toxic metals into the cell. The interaction of metal with the cell wall is a complex mechanism that involves several processes, such as ion exchange, complexation, adsorption and precipitation, including crystallization (Siddiquee et al. 2015).

The biosorption of toxic metal ions by microorganisms is an effective way to remove them from wastewater or soil. Most biosorption studies are conducted on fungi, as they can be grown easily and inexpensively (Malik et al. 2015). In addition, fungi have a high degree of resistance to the action of heavy metals and exhibit high sorption activity in relation to them. *A. niger* is actively used to study biosorption mechanisms (Tian et al. 2019).

Compared to other biosorption agents, fungal biomass has high percentage of cell wall materials that are involved in metal binding and which can take considerable quantities of heavy metals even in the absence of physiological activity (Siddiquee et al. 2015). The dead *A. niger* biomass efficiently removed different heavy metals (chromium (VI) and zinc (II) (100%), mercury (II) (83.2%), and fluorine (I) (83%)) at different pH conditions (4.0 for lead (II), 5.0 for zinc (II) and cobalt (II), and 5.5 for mercury (II)) (Acosta-Rodríguez et al. 2018). *A. niger* adsorbs cadmium, which is considered one of the most toxic heavy metals in large quantities and, therefore, can also be used to extract cadmium from waste and soil (Malik et al. 2015).

The release of organic acids by fungi immobilizes soluble metal ions or complexes in the form of insoluble compounds. The formation of complex compounds with organic acids reduces the activity and toxicity of free cations in soil or on a mineral substrate and gives a chance for the survival of fungi in an environment with a high content of toxic metals (Fomina et al. 2005; Munir et al. 2005; Chandran et al. 2014). Among low molecular weight organic acids, oxalic acid is the strongest binding agent, the interaction of which with metals can occur both extracellularly and within the cell. By forming stable complexes and/or insoluble salts (oxalates) with a number of metals, oxalic acid greatly reduces the bioavailability of metals. Therefore, oxalate formation is considered as a highly effective method of immobilization of many metals (Munir et al. 2005; Chandran et al. 2014).

The absorption of metals by fungi takes place as a result of transport systems operation. Metal transporters localized on membranes can uptake common nontoxic metals, but they can also pass other metals, including those that are undesirable for the cell. Once in the cell, the metal can be immobilized inside vacuoles or in the cytoplasm with specific proteins rich in SH peptide (metallothioneins and phytochelatins) (Siddiquee et al. 2015).

Oxalate-production of *A. niger* can be used not only for bioremediation. For example, oxalate films formed by microorganisms can be used for bioprotection of stone (Garcia-Valles et al. 1996; Mcllroy de la Rosa et al. 2012), primarily to counteract sulfatization, the formation of gypsum-enriched patina (Fassina et al. 1996; Maravelaki-Kalaitzaki 2005; Realini et al. 2005). Nevertheless, the real practical use of oxalate formation under the action of fungi for preserving stone due to the corrosive activity of fungal metabolites is problematic (Rampazzi 2019).

5.2. Strains Variability of *A. niger* in Resistance to Heavy Metals

In our study, we screened the of ten *A. niger* strains for their resistance to zinc and copper in a concentration range from 100 to 2000 μmol (Sazanova and Vlasov 2020). A detailed description of the strains is given in section 2 (Table 2.3). In relation to copper, strains isolated from the surface of a rock substrate, as well as antarctic strains showed the greatest esistance. Significant differences between the strains in resistance to zinc were not detected.

The increased resistance of strains isolated from the rock surface is probably associated with organic acids overproduction. The resistance of polar strains of microorganisms to heavy metals and other stress factors is associated with carbon metabolism, nitrogen metabolism, and protein folding processes (Khan et al. 2019). Obviously, the antarctic *A. niger* strains have a biochemical potential that allows them to adapt to adverse conditions, which makes them promising objects for further studies of fungal resistance to stress, developments in the field of bioremediation and biotechnology.

5.3. The Influence of Cultivation Conditions on the Resistance of *A. niger* to Heavy Metals

The properties of the environment are of great importance for the mechanisms of fungal resistance to toxic metals. According to our data, *A. niger* (strain Ch 4/07) interacts differently to zinc and copper metals at the physiological and biochemical level, depending on the composition of the nutrient medium (Sazanova et al. 2015). For the experiment, metal concentrations (inhibiting the growth of mycelium by 15-30%) 2 mM of the medium for Zn and 0.5 mM of the medium for Cu were selected. On Czapek medium with a nitrate form of nitrogen metals did not significantly affect the growth pattern of fungal hyphae at the micromorphological level. On Rollen nitrate-ammonium medium the addition of zinc and copper to the medium led to significant disturbances in the pattern of mycelium growth: at the micromorphological level, the addition of metals to the medium led to the formation of "swollen" cells, often making up the majority of the mycelium (Figure 5.1c-f).

The analysis of organic acids content in *A. niger* culture fluid showed an increase in oxalic acid production by 30-35% on a nitrate medium when exposed to zinc at a concentration of 2 mM (Table 5.1). At the same time, the formation of succinic and citric acids, which occurred in the control of *A. niger* in the presence of Zn, was completely suppressed.

When copper was added at a concentration of 0.5 mM on a nitrate medium, the content of oxalic acid in the cultural liquid, on the contrary, decreased significantly with respect to the control (from 57.8 to 2.2 μg/g), but at the same time, the release of succinic, malic, and, especially, citric acids increased. On a nitrate-ammonium medium under the action of zinc, there was no significant increase in the production of oxalic and other organic acids.

The study of the culture fluid showed that in *A. niger*, on the Czapek nitrate medium, oxalic acid in the control was present in two forms – soluble and insoluble in the ratio of 1.2:1.0. When zinc was added to this medium, almost all oxalic acid, the total amount of which increased, was in insoluble form (Table 5.2). When copper was introduced into the same medium, insoluble forms of oxalate in *A. niger* were practically not observed, and the amount of soluble oxalate turned out to be as low as on a medium with zinc.

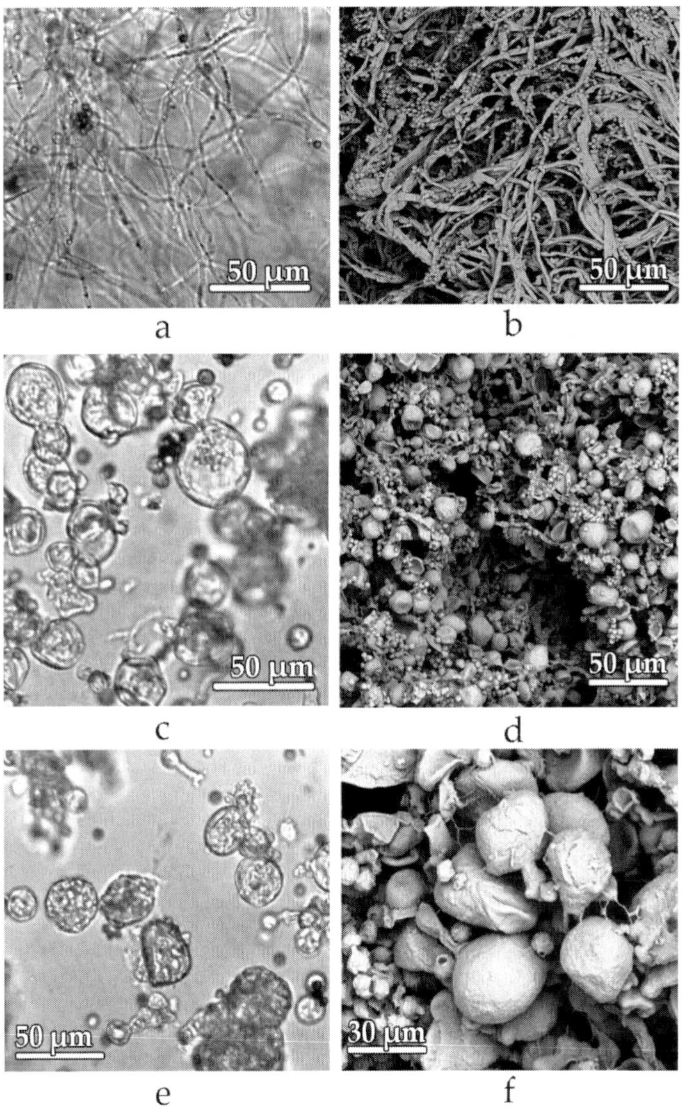

Figure 5.1. *A. niger* mycelium on Rollen nitrate-ammonium medium: a-b – control; c-d – with the addition of Cu 0.5mM; e-f – with the addition of Zn 2mM. a, c, e – optical microscopy; b, d, f – scanning electron microscopy.

Table 5.1. The influence of Zn and Cu on the production of organic acids by *A. niger* on liquid nutrient media with different nitrogen sources (10th day of growth)

Medium (nitrogen source)	Experiment	Final pH	Organic acid production, µg/g of micelium				
			oxalic	citric	succinic	malic	fumaric
Czapek ($NaNO_3$)	Control	2.75 ± 0.05	57.8 ± 7.8	5.5 ± 1.2	1.0 ± 0.2	-	-
	$ZnSO_4$ 2 mM	3.07 ± 0.04	89.6 ± 11.6	-	-	-	-
	$CuSO_4$ 0.5 mM	3.58 ± 0.02	2.2 ± 0.4	12.1 ± 1.1	2.1 ± 0.4	0.70 ± 0.03	-
Rollen (NH_4NO_3)	Control	1.69 ± 0.05	0.11 ± 0.02	6.6 ± 1.2	0.28 ± 0.03	0.45 ± 0.9	0.04 ± 0.04
	$ZnSO_4$ 2 mM	1.88 ± 0.06	0.16 ± 0.04	8.40 ± 1.20	0.29 ± 0.04	0.40 ± 0.07	0.04 ± 0.01
	$CuSO_4$ 0.5	1.71 ± 0.01	-	1.87 ± 0.56	0.24 ± 0.06	1.02 ± 0.20	-

During the cultivation of *A. niger* in the Czapek-Dox medium with the addition of zinc, the formation of zinc oxalate dihydrate intergrowths occurred on hyphae, as well as in the medium. On media with the addition of copper Cu-containing oxalates were not found.

Table 5.2. The content of different oxalate forms in *A. niger* culture fluid grown on Czapek-Dox culture medium

Experiment	Forms of calcium oxalate under the participation of *A. niger*	
	Unsoluble	Soluble
Control (no additives)	260.5 ± 32.2	312.1 ± 29.0
$ZnSO_4$ 2 mM	618.4 ± 28.1	12.2 ± 2.4
$CuSO_4$ 0.5 mM	1.3 ± 0.4	13.8 ± 3.4

As a result of quantitative determination of Zn and Cu content in fungal mycelium (for initial concentrations of 2 mM for zinc and 0.5 mM for copper in the medium) it was found both metals accumulate in the mycelium. In this case, Zn accumulates much more intensively in comparison with Cu. The level of zinc accumulation in the mycelium was three times higher during cultivation on nitrate than on ammonium nitrate media, whereas the opposite relationship was observed for copper (Table 5.3).

Table 5.3. The accumulation of metals in *A. niger* mycelium on various media

Nutrient medium (nitrogen source)	Metal concentration, μg/g of micelium	
	Zn	Cu
Czapek ($NaNO_3$)	4513 ± 80	55 ± 2
Rollen (NH_4NO_3)	1571 ± 33	353 ± 38

According to our data we can propose that, apparently, an active accumulation of metal ions inside the mycelium occurs on the ammonium nitrate medium, which leads to the formation of "swollen cells." The accumulation of zinc in large quantities in fungal mycelium on a nitrate medium is most likely due to its extracellular binding by excreted oxalic acid and the deposition of zinc oxalate crystals on the surface of the mycelium, which determines the decrease in Zn toxicity. Zn oxalates can form morphologically distinct intergrowths. Similar zinc oxalate intergrowths were found by Sayer and Gadd (2001) in *A. niger* culture, they were significantly smaller and morphologically different, but also had a lamellar. Zinc oxalate hydrates were also found in cultures of other fungi during their growth on Zn-containing media. All of these crystals had a lamellar form, but the shape of the intergrowths was very diverse (Gadd et al. 2007). There is evidence in published data that copper oxalates are also formed in fungal cultures (Fomina et al. 2005). Copper compounds with oxalic acid are stronger than complexes with C3 and C4 acids, however, they are rarely present in natural organic material. Copper compounds with other organic acids can also reduce their toxicity (Manceau and Matynia 2010). High concentration of copper inhibited the release of oxalic acid by fungi, but at the same time there is a reliable data that it stimulated the formation of malic and citric acids in them, which can effectively bind and reduce the toxicity of metals.

As in many other areas of bioremediation, many studies in the field of metal bioremediation with fungi are based on laboratory studies with little practical use. However, there are examples of large-scale bioremediation where microorganisms, including fungi, have been used to treat metal contaminated soil (Gadd 2000). Metals get into farm wastewater from animal feed and *A. niger* can remove 91% copper and 70% zinc from them (Price et al. 2001). Many fungal species have been reported to be used in the process of cleaning polluted areas such as *Trichoderma autroviride*, *T.harzianum*, *T.virens* and *A. niger* (Karcprzak and Malina 2005; Iskandar et al. 2011; Siddiquee et al. 2013). *A. niger* appears to be the most appropriate in the wastewater purification from metals by producing various extracellular proteins, organic acids and other metabolites (Chandran et al. 2014).

The accumulated data indicate that *A. niger* has a biochemical potential that allows us to consider it as a promising object for bioremediation and other biotechnologies. The role of organic acids and, in particular, oxalic acid in fungal resistance to heavy metals can be determined by the nature of the metal and its ability to form stable complexes with an acid anion, which reduces metal toxicity for fungi. The stimulating effect of metals on acid production largely depends on metal concentration, nitrogen form in a medium, and other cultivation conditions. Unfortunately, most bioremediation studies are conducted only in the laboratory. There are very few examples of their practical application.

CONCLUSION

The review of the results of modeling experiments on synthesizing different oxalate minerals by participation of the *A. niger* showed that the metabolism of rock inhabiting microfungi is an important factor of modern mineral formation. Mycogenic crystallization of oxalates can take place on the surface of different rocks and minerals: carbonates, phosphates, oxides, silicates and others. It should be kept in mind that fungi are part of a multicomponent lithobiotic community (bacteria, microalgae, lichens), the metabolism of which can differ significantly from fungal metabolism monoculture.

An insignificant number of oxalates found to date in biofilms on the surface of rocks and minerals is primarily due to the difficulty of their detection as they are not visible to the naked eye. In addition, the study of biofouling of stony substrates was carried out mainly by biologists who focused on the features of biological colonization, biodiversity and destructive activity of microorganisms, rather than biogenic crystallization. In the future, the number of biofilm minerals should increase significantly. In particular, we can expect the discovery of biogenic analogues of such minerals as iron oxalate (humboldtine), manganese oxalate (falottaite), copper oxalates (wheatleyite and fiemmeite), as well as discovering biominerals of zinc, lead and nickel oxalates. However it should be kept in mind that the interaction of various metabolic products can strongly affect the chemistry of the crystallization medium and inhibit the formation of oxalates even in the presence of active oxalic acid producers in the biofilm.

Morphogenetic patterns of the formation of oxalates in laboratory conditions under the influence of monocultures of fungi are affected by the chemistry of the crystallization medium, which is associated with both the fungal metabolic activity and the properties of underlying mineral substrate. The effect of the underlying substrate on the process of oxalate crystallization depends on its elemental and mineral composition, as well as structural features (porosity, density), i.e., from the parameters determining its solubility. The organic and inorganic components of the crystallization medium actively interact with the formed calcium oxalates, either selectively adsorbing on the faces of growing crystals, or through incorporation into oxalates, forming solid solutions.

The patterns revealed in model experiments involving *A. niger* on different mineral substrates should be used with caution when studying crystallization processes in natural biofilms. This is primarily due to the complex composition of the metabolites of the microbial community, as well as the significant environmental impact on the elemental composition of biofilms, in particular with selective accumulation of elements from the environment by different microorganisms (Frank-Kamenetskaya et al. 2019).

The accumulated data indicate that the intense metabolism of *A. niger*, in particular, its production of organic acids, allows us to consider it as a promising object in the development of biotechnologies, which should find wide practical application. The revealed wide range of ionic substitutions, common for metal oxalates formed on mineral substrates under the action of microorganisms, indicates the possibility of isolation of toxic elements from the environment in this way, i.e., opens a new promising approach for bioremediation of contaminated areas.

Acknowledgments

This work is supported by Russian Science Foundation (project N 19-17-00141). The laboratory researches were carried out in the Research Resource Centers of Saint Petersburg State University: SEM investigations

– in the center "Development of molecular and cell technologies" and "Microscopy and Microanalisys" and XRD measurements – in the center for X-ray Diffraction Studies.

References

Abín, L., Coto, O., Gómez Y., and Bosecker K. 2002. "Screening of fungi with capacity for organic acid production." *Revista Biologia* 16(1):69-73.

Acosta-Rodríguez, I., Cárdenas-González, J. F., Rodríguez Pérez, A. S., Oviedo, J. T., and Martínez-Juárez, V. M. 2018. "Bioremoval of Different Heavy Metals by the Resistant Fungal Strain *Aspergillus niger.*" *Bioinorg Chem Appl.*: 3457196. https://doi.org/10.1155/2018/3457196.

Adamo, P. and Violante, P. 2000. "Weathering of rocks and neogenesis of minerals associated with lichen activity." *Applied Clay Science* 16:229–256.

Adamo, P., Marchetiello, A. and Violante, P. 1993. "The weathering of mafic rocks by lichens." *Lichenologist* 25(3): 285-297.

Adamo, P., Colombo, C. and Violante, P. 1997. "Iron oxides and hydroxides in the weathering interface between Stereocaulon vesuvianum and volcanic rock." *Clay Minerals* 32: 453-461.

Adeyemi, A. O. 2009. "Biological Immobilization of lead from lead sulphide by *Aspergillus niger* and *Serpula himantioides.*" *International Journal of Environmental Research* 3(4):477-482.

Adeyemi, A. O. and Gadd, G. M. 2005 "Fungal degradation of calcium-, lead- and silicon-bearing minerals." *BioMetals* 18:269–281.

Ahmad, M. M., Ali, A., Khan, M. A. and Abdin, M. Z. 2015. "Biomolecular Characteristics of *Aspergillus niger* Under Cadmium Metal Stress." *Environmental Processes* 2(1):241–250.

Al-Shehri, A., and Mostafa, Y. S. 2006. "Citric acid production from date using syrop immobilized cells of *Aspergillus niger.*" *Biotechnology* 5(4):461-465.

Ali, S., Ikram-ul-H., Qadeer, M. A. and Javed, I. 2002a. "Production of citric acid by *Aspergillus niger* using cane molasses in a stirred fermentor." *Electronic Journal of Biotechnology* 5(3). Accepted November 5, 2002. doi: 10.2225/vol5-issue3-fulltext-3.

Ali, S., Ikram-ul-H., Javed, I.. 2002b. "The role of Mn++ ions for high and consistent yield of citric acid in recycling fed-batch bioreactor system and its novelty on kinetic basis." *Electron. J. Biotechnol.* 5(2):4-5. Accepted May 31, 2002. doi: 10.2225/vol5-issue2-fulltext-3.

Aliverdieva, D. A., Mamaev, D. M., Lagutina, L. S. and Bondarenko, D. I. 2010. "Transport of dicarboxylates in *Saccharomyces cerevisiae*." *Applied microbiology and microbial biotechnology*: 1611-1620.

Almousa, A. A., El-Ghany, A. M. N. and Ashour, E. H. 2018. "Citric Acid Fermentation by *Aspergillus niger*." *Journal of Innovations in Pharmaceutical and Biological Sciences* 5 (4): 20-37.

Anastassiadis, S. and Rehm, H-J. 2006. "Continuous gluconic acid production by Aureobasidium pullulans with and without biomass retention." *Electronic Journal of Biotechnology* 9(5). Accepted November 21, 2005. doi: 10.2225/vol9-issue5-fulltext-18.

Andersen, M. R., Salazar, M. P., Schaap, P. J., van de Vondervoort, P. J., Culley, D., Thykaer, J., Frisvad, J. C., Nielsen, K. F., Albang, R., Albermann, K., Berka, R. M., Braus, G. H., Braus-Stromeyer, S. A., Corrochano, L. M., Dai, Z., van Dijck, P. W., Hofmann, G., Lasure, L. L., Magnuson, J. K., Menke, H., Meijer, M., Meijer, S. L., Nielsen, J. B., Nielsen, M. L., van Ooyen, A. J., Pel, H. J., Poulsen, L., Samson, R. A., Stam, H., Tsang, A., van den Brink, J. M., Atkins, A., Aerts, A., Shapiro, H., Pangilinan, J., Salamov, A., Lou, Y., Lindquist, E., Lucas, S., Grimwood, J., Grigoriev, I. V., Kubicek, C. P., Martinez, D., van Peij, N. N., Roubos, J. A., Nielsen, J. and Baker, S. E. 2011. "Comparative genomics of citric-acid-producing *Aspergillus niger* ATCC 1015 versus enzyme-producing CBS 513.88." *Genome Res* 21(6): 885-897.

Ascaso, C., Galvan, J. and Rodgriguez-Pascual, C. 1982. "The weathering of calcareous rocks by lichens." *Pedobiologia* 24: 219-229.

Atencio, D., Coutinho, J. M. V., Graeser, S. b,, Matioli, P. A. and Menezes Filho, L. A. D. 2004. "Lindbergite, a new Mn oxalate dihydrate from Boca Rica mine, Galiléia, Minas Gerais, Brazil, and other occurences." *American Mineralogist* 89 (7): 1087-1091.

Balbontín, R., Vlamakis, H. and Kolter, R. 2014. "Mutualistic interaction between Salmonella enterica and *Aspergillus niger* and its effects on Zea mays colonization." *Microb Biotechnol.* 7(6): 589–600.

Baran, E. J. and Monje, P. V. 2008. "From Nature to Application, Metal Ions in Life Sciences." In *Biomineralization* edited by A. Sigel, H. Sigel, R. K. O. Sigel. Vol. 4, 219–254. Chichester*:* Wiley,

Baran, E. 2014. "Review: Natural oxalates and their analogous synthetic complexes." *Journal of Coordination Chemistry* 67(23–24): 3734–3768.

Barinova, K. V., Shchiparev, S. M., Shavarda, A. L. and Vlasov, D. Yu. 2010. "The influence of calcium carbonate on the acidifying activity of micromycetes." *Vestnik SPbGU* 3(3): 93-98. (In Russ).

Bennet-Klark, T. A. 1938. *The role of organic acids in plant metabolism.* Leningrad: State Publishing House for Biological and Medical Literature. (In Russian).

Benoit, I., van den Esker, M. H., Patyshakuliyeva, A. and Kovács, Á. T. 2015. "*Bacillus subtilis* attachment to *Aspergillus niger* hyphae results in mutually altered metabolism." *Environmental Microbiology* 17(6):2099-2113.

Bercovitz, A., Peleg, Y., Battat, E., Rokem, J. S. and Goldberg, I. 1990. "Localization of pyruvate carboxylase in organic acid producing *Aspergillus strains.*" *Appl. Environ. Microbiol.* 56: 1594–1597.

Berdoulay, M. and; Salvado, J. C. 2009."Genetic characterization of microbial communities living at the surface of building stones." *Letters in Applied Microbiology.*, 49: 311–316.

Betlej, I. and Grąz, M. 2006. "The identification of organic acid in Treametes versicolor culture, growing on a medium with CuHDO complex." *Folia forestalia polonica* 3:3-7.

Bjelland, T., Smbo, L. and Thorseth, I. H. 2002. "The occurrence of biomineralization products in four lichen species growing on sandstone in western Norway." *Lichenologist* 34(5): 429-440.

Bloom, S. J. and Johnson, M. J. 1962. "The pyruvate carboxylase of *Aspergillus niger*." *J. Biol. Chem*. 237: 2718–2720.

Bonaventura, M. P. D., Gallo, M. D., Cacchio, P., Ercole, C. and Lepidi, A. 1999 "Microbial Formation of Oxalate Films on Monument Surfaces Bioprotection or Biodeterioration?" *Geomicrobiology Journal.* 16:55-64.

Boonchan, S., Britz, M. L. and Stanley, G. A. 2000 "Degradation and mineralization of high-molecularweight polycyclic aromatic hydrocarbons by defined fungal-bacterial cocultures." *Applied and Environmental Microbiology*., 66:1007-1019.

Bohner, M. 2000. "Calcium orthophosphates in medicine: from ceramics to calcium phosphate cements." *Injury* 31(4): 37–47.

Braams, J. 1992. "*Ecological studies on the fungal microflora inhabiting historical sandstone monuments*". Thesis… Dr. rer. nat., Oldenburg University.

Bravo D. 2013 "*Assessing the diversity and metabolism of oxalotrophic bacteria in tropical soils*." Thesis… Dr. of Microbiology, University of Neuchâtel, Switzerland.

Britto, D. T. and Kronzucher, H. J. 2002. "NH_4^+ toxicity in higher plants: a critical review." *J. Plant Physiol.* 159: 567-284.

Bungartz, F., Garvie, L. A. J. and Nash, T. H. Anatomy of the endolithic Sonoran Desert lichen Verrucaria rubrocincta Breuss: implications for biodeterioration and biomineralization. *The Lichenologist*, 2004, 36 (1): 55–73

Burford, E. P., Kierans, M. and Gadd, J. M. Geomycology: fingi in mineral substrata. *Mycologist,* 2003, 17(3), 98-107.

Butkevich, V. S. 1957. *Selected Works*. Moscow: Publishing House of the Academy of Sciences. (In Russian).

Cai, L., Xiao, H.-R., Huang, S.-M., Li, H. and, Zhou, G.-T. 2013. "Solubilization of magnesium-bearing silicate minerals and the subsequent formation of glushinskite by *Aspergillus niger*." *Geomicrobiology Journal* 30: 302–312.

Cao, C., Jiang, J., Sun, H., Huang, Y., Tao, F. and Lian, B. 2016. "Carbonate Mineral Formation under the Influence of Limestone-Colonizing Actinobacteria: Morphology and Polymorphism." *Front Microbiol.* 7:366.

Carlile, M. J., Watkinson, S. C. and Gooday, G. W. 2001. *The Fungi*. San Diego: Academic press.

Carol, A., Engel, A., Straathof, A. J., Zijlmans, T. W., van Gulik, W. M. and van der Wielen, L. A. M. 2008. "Fumaric acid production by fermentation." *Applied Microbiology and Biotechnology* 78(3): 379–389.

Ceci, A., Kierans, M., Hillier, S., Persiani, A. M. and Gadd G. M. 2015. "Fungal bioweathering of mimetite and a general geomycological model for lead apatite mineral biotransformations." *Applied and Environmental Microbiology* 81:4955 – 4964.

Chandran, S. C, Shijith, K. V., Vipin K. V. and Augusthy, A. R. 2014. "Study on heavy metals toxicity biomarkers in *Aspergillus niger*." *IJAPBC* 3(2): 458-464.

Chisholm, J. E., Jones, G. C. and Purvis, O. W. 1987. "Hydrated copper oxalate, moolooite in lichens." *Mineralogical magazine* 51: 766-803.

Chen, J., Blume, H. P. and Beyer, L. 2000. "Weathering of rocks induced by lichen colonization – a review." *Catena* 39:121-146.

Costa, C., Rocha, S. M. and Almeida, A. 2014. "Metabolomics profile of *Aspergillus niger* through comprehensive two dimensional gas chromatography". *Trends in Environmental Microbiology for Public Health*, abstract book.

Costa, C. P., Silva, D. G., Rudnitskaya, A., Almeida, A., Rocha, S. M. 2016. "Shedding light on *Aspergillus niger* volatile exometabolome." *Scientific Reports* 6. Accepted 09 May 2016. doi: https://doi.org/10.1038/srep27441.

Demartin, F., Campostrini, I., Ferretti, P. and Rocchetti, I. 2018. "Fiemmeite $Cu_2(C_2O_4)(OH)_2\_2H_2O$, a New Mineral from Val di Fiemme, Trentino, Italy." *Minerals* 8: 248.

Dhami, N. K., Reddy, M. S. and Mukherjee A. 2013. "Biomineralization of Calcium Carbonate Polymorphs by the Bacterial Strains Isolated from Calcareous Sites." *Journal of microbiology and biotechnology*. 23:707-714.

Dhami, N. K., Reddy, M. S., Mukherjee, A. 2014. "Application of calcifying bacteria for remediation of stones and cultural heritages." *Front. Microbiol*. *5*:304.

Driessens, F. C. M. 1973. "Relation between apatite solubility and anti-cariogenic effect of fluoride." *Nature* 243(5407):420–421.

Elshafei A. M. 1989. "Degradation of some sugars and sugar acids by the nonphosphorylated D-gluconate pathway in Aspergillus ustus." *Acta Biotechnologica*. 9(5): 485 – 489.

Ercole. C., Cacchio. P., Botta. A. L., Centi. V., Lepidi. A. 2007. "Bacterially Induced Mineralization of Calcium Carbonate: The Role of Exopolysaccharides and Capsular Polysaccharides." *Microscopy and Microanalysis*. 13:42-50.

Ehrlich, H. L. 1999. "Microbes as geologic agents: their role in mineral formation." *Geomicrobiology Journal* 16:135-153.

Fassina, V., Rossetti, M. and Rattazzi, A. 1996. "Oxalate films on Venetian monuments using different analytical techniques to investigate the mechanism of formation." In *Proceedings of II International Symposium 'The Oxalate Films in the Conservation of Works of Art,* edited by L. Toniolo, M. Realini, 299–315. Italy.

Ferrier, J., Yang, Y., Cseteny., L. and Gadd, G. M. 2019. "Colonization, penetration and transformation of manganese oxide nodules by *Aspergillus niger*." *Environmental Microbiology* 21(5):1821-183.

Fomina, M., Hillier, S., Charnock, J. M., Melville, K., Alexander, I. J. and Gadd, G. M. 2005. "Role of oxalic acid overexcretion in transformations of toxic metal minerals by *Beauveria caledonica*." *Applied and Environmental Microbiology* 71(1): 371-381.

Foster, D. 1950. *Chemical activity of mushrooms*. Moscow: publishing house of foreign literature. (In Russian).

Frank-Kamenetskaya, O. V., Ivanyuk, G. Y., Zelenskaya, M. S., Izatulina, A. R., Kalashnikov, A. O., Vlasov, D. J. and Polyanskaya, E. I. 2019.

"Calcium Oxalates in Lichens on Surface of Apatite-Nepheline Ore (Kola Peninsula, Russia)." *Minerals* 9: 656.

Frank-Kamenetskaya, O. V., Vlasov, D. Y. and Rytikova, V. V. 2019. *The Effect of the Environment on Saint Petersburg's Cultural Heritage: Results of Monitoring the Historical Necropolis Monuments*. Berlin: Springer.

Frank-Kamenetskaya, O. V., Vlasov, D. Yu. and Shilova O. A. 2012. "Biogenic Crystals Genesis on a Carbonate Rock Monument Surface: The Main Factors and Mechanisms, the Development of Nanotechnological Ways of Inhibition." In *Minerals as advanced Materials II.* Edited by Krivovichev, S., 401-413. Springer –Verlag: Berlin Heidelberg.

Gadd, G. M. 1993. "Interactions of fungi with toxic metals." *New Phytol* 124: 25-60.

Gadd, G. M. 1996. "Influence of microorganisms on the environmental fate of radionuclides." *Endeavour*. 20:150–156.

Gadd, G. M. 1997. "Roles of microorganisms in the environmental fate of radionuclides. "In *CIBA Foundation Symposia No. 203: Health Impacts of Large Releases of Radionuclides*, edited by Lake, J. V., Bock G. R. & G. Cardew, 94–108. Chichester, UK: Wiley & Sons.

Gadd, G. M. 1999. "Fungal production of citric and oxalic acid: Importance in metal speciation, physiology and biogeochemical processes" *Adv. Microb. Physiol*. 41:47-91.

Gadd, G. M. 2000. "Bioremedial potential of microbial mechanisms of metal mobilization and immobilization." *Current Opinion in Biotechnology*, 11: 271–279.

Gadd, G. M. 2001. *Fungi in Bioremediation*. Cambridge: Cambridge University Press.

Gadd, M. G. 2004. "Mycotransformation of organic and inorganic substrates." *Mycologist* 18(2): 60-70.

Gadd, G. M. 2007. "Geomycology: biogeochemical transformations of rocks, minerals, metals and radionuclides by fungi, bioweathering and bioremediation." *Mycological research*. 111:3 – 49.

Gadd, M. G. 2017. "Fungi, rocks and minerals." *Elements* 13: 171-176.

Gadd, G. M. and Sayer, G. M. 2000. “Fungal transformations of metals and metalloids.” In *Environmental Microbe–Metal Interactions*, edited by D. R. Lovley, 237– 256. Washington, DC: American Society for Microbiology.

Gadd, G. M., Bahri-Esfahani, J., Li, Q., Rhee, Y. J., Wei, Z., Fomina, M. and Liang, X. 2014. “Oxalate production by fungi: significance in geomycology, biodeterioration and bioremediation.” *Fungal Biological Reviews* 28: 36–55.

Gallmetzer, M. and Burgstaller, W. 2002. “Efflux of organic acids in Penicillium simplicissimum is an energy-spilling process, adjusting the catabolic carbon flow to the nutrient supply and the activity of catabolic pathways.” *Microbiology* 148: 1143–1149.

Garcia-Valles, M., Blazquez, F., Molera, J. and Vendrell-Saz, M. 1996 “Studies of patinas and decay mechanisms leading to the restoration of Santa Maria de Montblanc (Catalonia, Spain).” *Stud. Conserv.* 41: 1–8.

Gehrmann, C. K. and Krumbein, W. E. 1994. “Interaction between epilithicans endolithic lichens and carbonate rocks.” *In III International symposium on the Conservation of Monuments in the Mediterranean Basin*, 311-316.

Gharieb, M., Sayer, J. A., Gadd, M. G. Solubilization of natural gypsum ($CaSO_4.2H_2O$) and the formation of calcium oxalate by *Aspergillus niger* and *Serpula himantioides*. *Mycological Research,* 1998, V. 102, N.7, 825-830.

Ghorbani, Y., Oliazadeh, M., Shahvedi, A., Roohi, R. and Pirayehgar, A. 2007. “Use of some isolated fungi in biological Leaching of Aluminum from low grade bauxite.” *African Journal of Biotechnology* 6(11): 1284-1288.

Giester, G. 1997. “Syntheses and crystal structures of $Co_3(C_2O_4)(SeO_3)_2$ and $Zn(C_2O_4) \cdot 2H_2O$.” *Zeitschrift fur Kristallographie.* 10:720-723.

Gibson, B. R. and Mitchell, D. T. 2004. “Nutritional influences on the solubilization of metal phosphate by ericoid mycorrhizal fungi.” *Mycologycol Research* 108(8): 947–954.

Gilbert, P., Maira-Litran, T., McBain, A. J., Rickard, A. H., Whyte, F. W. 2002 "The physiology and collective recalcitrance of microbial biofilm communities." *Advances in microbial physiology.*, 46:203-256.

Gorbushina, A. A. Life on the rocks. *Environmental Microbiology*, 2007, V.9, N.7, 1613-31.

Graeser, S. and Gabriel, W. 2013. "Falottaite, IMA 2013-044. CNMNC Newsletter No. 17, October 2013, page 3000." *Mineralogical Magazine* 77: 299-3005.

Graeser, G. and Gabriel, W. 2016. "Falottait $(MnC_2O_4)\cdot 3H_2O)$ - ein neues Oxalat-Mineral aus den Schweizer Alpen." *Schweizer Strahler* 50(3): 20-27.

Han Y., Joosten H. J., Niu W., Zhao Z., Mariano P. S., Calman M. M., van Kan J., Schaap P. J. and Dunaway-Mariano D. 2007. "Oxaloacetate hydrolase, the C–C bond lyase of oxalate secreting fungi." *The journal of biological chemistry* 282(13): 9581–9590.

Hayaishi, O., Shimazo, H., Katagiri, M. and Satio, Y. 1956. "Enzymatic formation of oxalate and acetate from oxaloacetate." *J. Am. Chem. Soc.* 8: 5126-5127.

Hess, C. D., Lu, W., Rabinowitz, D. J. and Botstein, D. 2006. "Ammonium toxicity and potassium limitation in yeast." *PLoS Biology* 4(11): 2012-2023.

Holligan, P. M. and Jennings, D. H. 1972. "Carbohydrate metabolism in the fungus Dendryphiella Salina." *New Phytol* 71: 1119-1133.

Hoppert, M., Flies, C., Pohl, W., Gunzl, B. and Schneider, J. 2004. "Colonization strategies of lithobiontic microorganisms on carbonate rocks." *Environmental Geology* 46:421-428.

Hossain, M., Brooks, J. D. and Maddox, I. S. 1984. "The effect of the sugar source on citric acid production by *Aspergillus niger*." *Appl Microbiol. Biotechnol.* 19 (I.6): 393–397.

Iskandar, N. L., Zainudin, N. A. and Tan, S. G. 2011. "Tolerance and biosorption of copper (Cu) and lead (Pb) by filamentous fungi isolated from a freshwater ecosystem." *J Environ Sci* 23: 824-830.

Izatulina, A. R., Gurzhiy, V. V., Krzhizhanovskaya, M. G., Kuz'mina, M. A., Leoni, M. and Frank-Kamenetskaya, O. V. 2018. "Hydrated

Calcium Oxalates: Crystal Structures, Thermal Stability and Phase Evolution." *Cryst. Growth Des.* 18: 5465–5478.

Jenni, B., Aragno, M. and Wiegel, J. K. W. 1987. "Numerical analysis and DNA-DNA hybridization studies on Xanthobacter and emendation of Xanthobacter flavus." *Systematic and Applied Microbiology*. 9. 247–53.

Jenni, B., Realini, L., Aragno, M. and Tamer, A. U. 1988. "Taxonomy of non H2-lithotrophic, oxalate-oxidizing bacteria related to Alcaligenes eutrophus." *Systematic and Applied Microbiology*. 10. 126–30.

Jennings, D. H. 2007. *The physiology of fungal nutrition*. Cambridge: Cambridge University Press.

Jones, D., Wilson, M. J., Tait, J. M. Weathering of a basalt by Pertusaria corallina. *Lichenologist*, 1980, 12 (3): 277-289.

Joseph, J. F. L. and van Heijnen, W. M. G. 2015. "Development of tools for quantitative intracellular metabolomics of *Aspergillus niger* chemostat cultures." *Metabolomics* 11(5):1253–1264.

Kang, X., Csetenyi, L. and Gadd, G. M. 2019. "Biotransformation of lanthanum by *Aspergillus niger*." *Applied Microbiology and Biotechnology* 103: 981–993.

Karcprzak, M. and Malina, G. 2005. "The tolerance and Zn2+, Ba2+ and Fe2+ accumulation by *Trichoderma atroviride* and *Mortierella exigua* isolated from contaminated soil." *Can J Soil Sci* 85: 283-290.

Kawai, H., Isobel, Y., Horibe, M., Tokuda, L., Tokuno, I., Endo, K. and Kawai, F. 1992. "Production of a Novel Extracellular Polysaccharide by a *Bacillus* Strain Isolated from Soil." *Biosci. Biotech. Biochem.*, 56: 853-857.

Keshari, N., Prasad, S. and Biofouling. A. 2013. "Characterization of cyanobacteria isolated from biofilms on stone monuments at Santiniketan, India." *The Journal of Bioadhesion and Biofilm Research.*, 29(5): 525–536.

Khan, I., Aftab, M., Shakir, S. et al. 2019. "Mycoremediation of heavy metal (Cd and Cr)–polluted soil through indigenous metallotolerant fungal isolates." *Environ Monit Assess* 191: 585. https://doi.org/10.1007/s10661-019-7769-5.

Khattab, A. A., Salem, A. A. and Soheam, A. E. 2017. "Optimization of citric acid production by *Aspergillus niger* isolated from different habitats." *Res. J. Pharmaceutic. Biol. Chem. Sci.* 8:614.

Kolo, K. and Claeys, Ph. 2005. "*In vitro* formation of Ca-oxalates and the mineral glushinskite by fungal interaction with carbonate substrates and seawater." *Biogeosciences Discuss* 2: 451-497.

Kretovich, V. L. 1980. *Plant biochemistry*. Moscow: Vyshcha Shkola. (In Russian).

Kubicek, C. P., Schreferl-Kunar, G., Wöhrer, W. and Röhr, M. 1988. "Evidence for a cytoplasmic pathway of oxalate biosynthesis in *Aspergillus niger*." *Appl Environ Microbiol*. 54: 633–637.

Kurakov, A. V., Somova, N. G. and Ivanovskii, R. N. 1999. "Micromycetes populating limestone and red brick surfaces of the Novodevichy Convent masonry." *Microbiology* 68 (2): 273-282. (In Russian).

Kuz'mina, M. A., Rusakov, A. V., Frank-Kamenetskaya, O. V. and Vlasov, D. Yu. 2019. "The influence of inorganic and organic components of biofilm with microscopic fungi on the phase composition and morphology of crystallizing calcium oxalates." *Crystallogr Rep* 64(1):161–167.

Lapeyrie, F., Chilvers, G. A. and Bhem, C. A. 1991. "Oxalic acid synthesis by myccorrhyzal fungus Paxillus involutus (Batsch. Ex. Fr.)." *Fr. New Phytol*. 1(106): 139-146.

Lee, E., Jeong, B. C., Park, Y. H. and Kim, H. H. 2014. "Expression of the gene encoding oxalate decarboxylase from *Bacillus subtilis* and characterization of the recombinant enzyme." *BMC Res Notes*. Sep 3. 7:598. doi: 10.1186/1756-0500-7-598.

Lenz, H., Wunderwald, P. and Eggerer, H. 1976. "Partial purification and some properties of oxaloacetase from *Aspergillus niger*." *Eur. J. Biochem*. 65: 225-235.

Ma, H., Kubicek, C. P. and Rohr, M. 1981. "Malate dehydrogenase isoenzymes from *Aspergillus niger*." *FEMS Microbiol. Lett* 12:147-151.

Magnuson, J. K. and Lasure, L. L. 2004. "Organic acid production by filamentous fungi." In *Advances in fungal biotechnology for industry, agriculture, and medicine*, edited by Jan and Lene Lange, 307-340. Kluwer Academic/Plenum Publishers.

Magyarosy, A., Laidlaw, R. D., Kilaas, R., Echer, C., Clark, D. S. and Keasling, J. D. "Nickel accumulation and nickel oxalate precipitation by *Aspergillus niger*." *Appl Microbiol Biotechnol*, 2002, 59:382–388.

Malik, D. S., Jain, C. K., Yadav, A. K. "Removal of heavy metals from emerging cellulosic low-costadsorbents: a review." *Appl Water Sci.* doi: 10.1007/s13201-016-0401-8

Manceau, A. 2010 "The nature of Cu bonding to natural organic matter" *Geochimica et Cosmochimica Acta.* 74: 2556–2580.

Maravelaki-Kalaitzaki, P. 2005. "Black crusts and patinas on Pentelic marble from the Parthenon and Erechtheum (Acropolis, Athens): characterization and origin." *Anal. Chim. Acta* 532:187–198.

Marques, J., Gonçalves, J., Oliveira, C., Favero-Longo, S. E., Paz-Bermúdez, G., Almeida, R. and Prieto, B. 2016. "On the dual nature of lichen-induced rock surface weathering in contrasting micro-environments." *Ecology* 97 (10): 2844–2857.

Marvasi, M., Visscher P. T. and Martinez L. C. 2010 "Exopolymeric substances (EPS) from *Bacillus subtilis*: polymers and genes encoding their synthesis" *FEMS Microbiology Letters.* 313(1):1–9.

Max, B., Salgado, J. M., Rodríguez, N., Cortés, S., Converti, A., and Domínguez, J. M. 2010. "Biotechnological production of citric acid." *Braz J Microbiol* 41(4): 862–875.

McIlroy de la Rosa, J. P., Warke, P. A. and Smith, B. J. 2012. "Lichen-induced biomodification of calcareous surfaces: Bioprotection versus biodeterioration." *Progr. Phys. Geography* 37(3):325–351.

Meti, S. R., Ambarish, S., Khajure, P. V. 2011. "Enzymes of ammonia assimilation in fungi: an overview." *Recent Research in Science and Technology* 2(4): 28-38.

Mills, S. J. and Christy, A. G. 2016. "The Great Barrier Reef Expedition 1928–29: The crystal structure and occurrence of weddellite, ideally

CaC2O42.5H2O, from the Low Isles, Queensland." *Mineral. Mag.* 80: 399–406.

Monje, P. V. and Baran, E. J. 2004. In *Advances in Plant Physiology* edited by A. Hemantaranjan. Jodhpur: Scientific Publishers 7: 395–410.

Müller, H. M. 1986. "Utilization of gluconate by *Aspergillus niger*. II. Enzymes of degradation pathways and main end products." *Zentralbl. Mikrobiol* 141(6): 461-469.

Munir, E., Yoon, J. J., Tokimatsu, T., Hattori, T. and Shimada, M. 2001. "New role for glyoxylate cycle enzymes in wood-rotting basidiomycetes in relation to biosynthesis of oxalic acid." *J. Wood Sci.* 47: 368–373.

Munir, E., Hattori, T. and Shimada, M. 2005. "Role for oxalate acid biosynthesis in growth of copper tolerant wood-rotting and pathogenic fungi under environmental stress." The *55th meeting of the Japan wood research society*. 1-7.

Muynck, W. D., Belie, N. D. and Verstraete, W. 2010. "Microbial carbonate precipitation in construction materials: A review." *Ecol. Eng*. 36.118-136.

O'Connor, B. H., Clarke, R. M. and Kimpton, J. A. 2019. "Synchrotron radiation diffraction study of the mineral moolooite, and synthetic copper oxalates." *Powder Diffraction,* 34(1): 21-34.

Odoni, D. I., van Gaal, M. P., Schonewille, T., Tamayo-Ramos, J. A., dos Santos, V. A. P. M., Suarez-Diez, M., and Schaap, P. J. "*Aspergillus niger* Secretes Citrate to Increase Iron Bioavailability." *Front. Microbiol.* 8:1424. doi.org/10.3389/fmicb.2017.01424.

Ozdal, M. and Kurbanoglu, E. B. 2019 "Citric Acid Production by *Aspergillus niger* from Agro-Industrial By-Products: Molasses and Chicken Feather Peptone." *Waste and Biomass Valorization* 10: 631–640.

Papagianni, M. 2007. "Advances in citric acid fermentation by *Aspergillus niger*: Biochemical aspects, membrane transport and modeling." *Biotechnology Advances* 25: 244–263.

Paris, F., Botton, B., Lapeyrie, F. 1996 "In vitro weathering of phlogopite by ectomycorrhizal fungi II. Effect of K^+ and Mg^{2+} deficiency and N source on accumulation of oxalate and H^+." *Plant and Soil.* 179: 141-150.

Pedersen, H., Christensen, B., Hjort, C. and Nielsen, J. 2000. "Cloning and characterization of oah, the gene encoding oxaloacetate hydrolase in *Aspergillus niger.*" *Mol. Gen. Genet.* 263: 281–286.

Peksel, A., Torres, N. V., Liu, J., Juneau, G. and Kubicek, C. P. 2002. "13C-NMR Analysis of glucose metabolism during citric acid production by *Aspergillus niger.*" *Appl. Microbiol. Biotechnol.* 58: 157–163.

Pel, J. H., de Winde, J. H., Archer, D. B., Dyer, P. S., Hofmann,G., Schaap, P. J., Turner, G., de Vries, R. P., Albang, R., Albermann, K., Andersen, M. R., Bendtsen, J. D., Benen, J. A. E., van den Berg, M., Breestraat, S., Caddick, M. X., Contreras, R., Cornell, M., Coutinho, P. M., Danchin, E. G. J., Debets, A. J. M., Dekker, P., van Dijck, P. W. M., van Dijk, A., Dijkhuizen, L., Driessen, A. J. M., d'Enfert, C., Geysens, S., Goosen, C., Groot, G. S. P., de Groot, P. W. J., Guillemette, T., Henrissat, B., Herweijer, M., van den Hombergh, J. P. T. W., van den Hondel C. A. M. J. J., van der Heijden, R. T. J. M., van der Kaaij, R. M., Klis, F. M., Kools, H. J., Kubicek, C. P., van Kuyk, P. A., Lauber, J., Lu, X., van der Maarel, M. J. E. C., Meulenberg, R., Menke, H., Mortimer, M. A., Nielsen, J., Oliver, S. G., Olsthoorn, M., Pal K., van Peij, N. N. M. E., Ram, A. F. J., Rinas, U., Roubos, J. A., Sagt, C. M. J., Schmoll, M., Sun, J., Ussery, D., Varga, J., Vervecken, W., van de Vondervoort, P. J. J., Wedler, H., Wösten, H. A. B., Zeng, A,-P., van Ooyen, A. J. J., Visser, J. and Stam, H. 2007. "Genome sequencing and analysis of the versatile cell factory *Aspergillus niger* CBS 513.88." *Natur biotechnology* 25(2): 221-231.

Pera, L. M., and Callieri, D. A. 1999. "Influence of calcium on fungal growth and citric acid production during fermentation of a sugarcane molasses-based medium by a strain of *Aspergillus niger*" *World Journal of Microbiology and Biotechnology* 15(I. 5.): 647-649.

Pinzari, F., Tate, J., Bicchieri, M., Rhee, Y. J. and Gadd, G. M. Biodegradation of ivory (natural apatite): possible involvement of fungal activity in biodeterioration of Lewis Chessman. *Environmental Microbiology*, 2013, 15(4):1050–1062.

Price, M. S., Classen, J. J. and Payne, G. A. 2001. "*Aspergillus niger* absorbs copper and zinc from swine wastewater." *Bioresour Technol. Mar.* 77(1):41-49.

Prieto, B., Silva, B., Rivas, T., Wierzchos, J. and Ascaso, C. 1997. Mineralogical transformation and neoformation in granite caused by the lichens Tephromela atra and Ochrolechia parella. *International Biodeterioration and Biodegradation* 40(2-4): 191-199.

Punin, Yu. O., Shtukenberg, A. G. 2008. *Autodeformational defects of crystals*. Saint Petersburg: St. Petersburg State University.

Purvis, O. W. 1984 "The occurrence of copper oxalates in lichens crowing on copper sulphide-bearing rocks in Scandinavia." *Lichenologist* 16(2): 197-204.

Purvis, O. W., Pawlik-Skowronska, B., Cressey, G., Jones, G. C., Kearsley, A. and Spratt, J. 2008. "Mineral phase and element composition of the copper hyperaccumularor lichen Lecanora polytropa." *Mineralogical Magazine* 72(2): 607-616.

Purvis, O. W., Kearsley, A. and Cressey, G. 2008. "Mineralization in rust-coloured Acarospora." *Geomicrobiology Journal* 25: 142-148.

Ramachandran, S., Fontanille, P., Pandey, A. and Larroche, C. 2006. "Gluconic acid: a review." *Food Technol. Biotechnol.* 44(2): 185–195.

Rampazzi L. 2019. "Calcium oxalate films on works of art: A review." *Journal of Cultural Heritage* https://doi.org/10.1016/j.culher.2019.03.002

Realini, M., Colombo, C., Sansonetti, A., Rampazzi, L., Colombini, M. P., Bona- duce, I., Zanardini, E. and Abbruscato, P. 2005 "Oxalate films and red stains on Carrara Marble." *Annal. Chim.* 95:217–226.

Ríos de los, A., Cámara, B., Cura del, M. Á. G., Rico, V. J., Galván, V. and Ascaso, C. 2009. "Deteriorating effects of lichen and microbial colonization of carbonate building rocks in the Romanesque churches of Segovia (Spain)." *Science of the Total Environment* 407: 1123–1134.

Rosling, A., Lindahl, B. D., Taylor, A. F. S. and Finlay, R. D. 2004. "Mycelial growth and substrate acidic cation of ectomycorrhizal fungi in response to different minerals." *Microbiology Ecology* 47: 31-37.

Rouse, R. C., Peacor, D. R., Dunn, P. J., Simmons, W. B. and Newbury, D. 1986. "Wheatleyite, $Na_2Cu(C_2O_4)_2 \cdot 2H_2O$, a natural sodium copper salt of oxalic acid." *American Mineralogist* 71: 1240-1242.

Ruijter, G. J., Panneman, H., van den Broeck, H. C., Bennett, J. M. and Visser, J. 1996. "Characterisation of the Aspergillus nidulans fra1 mutant: hexose phosphorylation and apparent lack of involvement of hexokinase in glucose repression." *FEMS Microbiol Lett* 139: 223–228.

Ruijter G. J. G., van de Vondervoort P. J. I., and Visser J. 1999. "Oxalic acid production by *Aspergillus niger*: an oxalate-non-producing mutant producescitric acid at pH 5 and in the presence of manganese." *Microbiology* 145: 2569–2576.

Rusakov, A. V., Frank-Kamenetskaya, O. V., Zelenskaya, M. S., Vlasov, D. Yu., Gimelbrant, D. E., Knauf, I. V. and Plotkina, Yu. V. 2010. "Calcium oxalates in bio-films on surface of the chersonesus archaeological limestone monuments (Crimea)." *Zapiski RMO (Proceedings of the Russian Mineralogical Society* 5: 96-104. (in Russian).

Rusakov, A. V., Vlasov, A. D., Zelenskaya, M. S., Frank-Kamenetskaya, O. V. and Vlasov, D. Yu. 2016. "The crystallization of calcium oxalate hydrates formed by interaction between microorganisms and minerals." In *Biogenic—Abiogenic Interactions in Natural and Anthropogenic Systems* edited by Frank-Kamenetskaya, O. V.; Panova, E. G., Vlasov, D. Yu. 357-377. Berlin, Germany: Springer Verlag.

Rusakov, A. V., Kuzmina, M. A., Izatulina, A. R. and Frank-Kamenetskaya, O. V. 2019. "Synthesis and Characterization of $(Ca,Sr)[C_2O_4]\_nH_2O$ Solid Solutions: Variations of Phase Composition, Crystal Morphologies and in Ionic Substitutions." *Crystals* 9: 654.

Sahin, N. 2003. "Oxalotrophic bacteria." *Research in Microbiology*. 154. 399–407.

Salvadori, O. and Municchia, A. C. The role of fungi and lichens in the biodeterioration of stone monuments. *The Open Conference Proceeding Journal*, 2015, 6 (Suppl 1:M4) 70-82.

Sarret, G., Manceau, A., Cuny, D., Haluwyn van, C., Deruelle, S., Hazemann, J. L., Soldo, I., Eybert-Berard, L. and Menthonnex, J. J. 1998. "Mechanisms of lichen resistance to metallic pollution." *Environ. Sci. Technol.* 32: 3325-3330.

Sayer, J. A., Cotter-Howells, J. D., Watson, C., Hillier, S. and Gadd, G. M. 1999. "Lead mineral transformation by fungi." *Current Biology*, 9: 691–694.

Sayer, J. A. and Gadd, J. M. 1997. "Solubilization and transformation of insoluble inorganic metal compounds to insoluble metal oxalates by *Aspergillus niger*." *Mycological Research* 6: 653-661.

Sayer, J. A. and Gadd, M. G. 2001. "Binding of cobalt and zinc by organic acids and culture filtrates of *Aspergillus niger* grown in the absence or presence of insoluble cobalt or zinc phosphate." *Mycological Research* 105 (10): 1261-1267.

Sayer, J. A., Kierans, M. and Gadd J. M. 1997. "Solubilisation of some naturally occurring metal-bearing minerals, limescale and lead phosphate by *Aspergillus niger*." *FEMS Microbiology Letters* 154: 29-35.

Sazanova, K. V., Frank-Kamenetskaya, O. V., Vlasov, D. Yu., Zelenskaya, M. S., Vlasov, A. D., Rusakov, A. V. and Petrova M. A. "Crystallization of calcium carbonates and oxalates on marble surface induced by metabolism of bacteria and bacterial-fungal associations." *Cryst. Growth Des. In press.*

Sazanova, K. V. and Vlasov D. Yu. 2020."Change of acid-production activity of *Aspergillus niger* (tiegh.) strains in adaptation to stress" *Mycology and phytopathology*. In press. (In Russian).

Sazanova, K., Osmolovskaya, N., Schiparev, S, Yakkonen, K., Kuchaeva, L. and Vlasov, D. 2015. "Organic Acids Induce Tolerance to Zinc- and Copper-Exposed Fungi Under Various Growth Conditions." *Current Microbiology*. 70: 520–527.

Sazanova, K. V., Vlasov, D. Yu., Osmolovskay, N. G., Schiparev, S. M. and Rusakov, A. V. 2016. "Significance and regulation of acids production by rock-inhabited fungi." In *Biogenic-Abiogenic Interactions in Natural and Anthropogenic Systems* edited by Frank-Kamenetskaya, O. V., Panova, E. G., Vlasov, D. Yu. 379-392. Switzerland: Springer International Publishing.

Schneider K., Skovran E. and Vorholta J. A. 2012. "Oxalyl-Coenzyme A Reduction to Glyoxylate Is the Preferred Route of Oxalate Assimilation in Methylobacterium extorquens" *J. Bacteriol.* 194(12): 3144-3155. doi: 10.1128/JB.00288-12.

Show, P. L., Oladele, K. O., Siew, Q. Y., Zakry, F. A. A., Lan, J. C.-W. and Ling T. C. 2015. "Overview of citric acid production from *Aspergillus niger.*" *Frontiers in Life Science Volume* 8(3) doi: org/10.1080/21553769.2015.1033653.

Siddiquee, S., Aishah, S. N., Azad S. A., Shafawati, S. N. and Naher L. 2013. "Tolerance and Biosorption capacity of Zn2+, Pb2+, Ni3+ and Cu2+ by filamentous fungi (*Trichoderma harzianum, T. aureoviride* and *T. virens*)." *Adv Biosci Biotechnol*. 4: 570-583.

Siddiquee, S., Rovina K., Azad S., Naher L., Suryani S. and Chaikaew P. 2015. "Heavy Metal Contaminants Removal from Wastewater Using the Potential Filamentous Fungi Biomass: A Review." *J Microb Biochem Technol*.7:6. doi: 10.4172/1948-5948.1000243.

Singh, O. V., Sharma, A. and Singh, R. P. 2001. "Optimisation of fermentation conditions for gluconic acid production by a mutant of *Aspergillus niger.*" *Indian J Exp Biol*. 39(11): 1136–1143.

Soldatenkov, S. V. 1971. *Biochemistry of Organic Plant Acids*. Leningrad: Publishing house of Leningrad University. (In Russian).

Somova, N. G., Dobrovol'skaya, T. G., Zenova, G. M. and Ivanovskii, R. N. 1998. "Microbial growth of the masonry surface: synecological analysis." *Microbiology* 67(5): 687-693 (*In Russian*).

Souza-Egipsy, V., Wierzchos, J., Carcia-Ramos, J. V. and Ascaso, C. 2002. "Chemical and ultrastructural features of the lichen-volcanic sedimentary rock interface in a semiarid region (Almeria, Spain)." *Lichenologist* 34 (2): 155-167.

Sterflinger, K. 2000. "Fungi as geological agents." *Geomicrobiology Journal* 17:97-124.

Sterling, C. 1965. "Crystal-structure analysis of weddellite." *Acta Crystallogr* 18: 917–921.

Sturm, E. V., Frank-Kamenetskaya, O., Vlasov, D., Zelenskaya, M., Sazanova, K., Rusakov, A. and; Kniep, R. 2015 "Crystallization of calcium oxalate hydrates by interaction of calcite marble with fungus *Aspergillus niger*." *Am. Mineral.*, 100: 2559–2565.

Sutjaritvorakul, T., Whalley, A. J. S. and Sihanonth, P. 2015. "Biotransformation of insoluble lead compounds by fungi isolated from polluted soil." *KMITL Sci. Tech. J.* 15 (2) 55-61.

Sutjaritvorakul, T., Gadd, M. G., Suntornvongsagul, K., Whalley, A. J. S., Roengsumran, S. and Sihanonth, P. 2013. "Sulubilization and transformation of insoluble zinc compounds by fungi isolated from a zinc mine." *Environment Asia* 6(2): 42-46.

Syers, J. K., Birnie, A. C. and Mitchell, B. B. 1967. "The calcium oxalate content of some lichens growing on limestone." *Lichenologist* 3: 409-414.

Tanner, A. and Bornemann, S. 2000. "*Bacillus subtilis* YvrK Is an Acid-Induced Oxalate Decarboxylase. *J Bacteriol. Sep.* 182(18): 5271–5273.

Tian, D., Jiang, Z., Jiang, L., Su, M., Feng, Z., Zhang, L., Wang, S., Li, Z. and Hu, S. 2019. "A new insight into lead (II) tolerance of environmental fungi based on a study of *Aspergillus niger* and Penicillium oxalicum." *Environmental Microbiology* 21(1):471-479.

Thomas, A., Rosseeva, E., Hochrein, O., Carrillo-Cabrera, W., Simon, P., Duchstein, P., Zahn, D. and Kniep, R. 2012. "Mimicking the growth of a pathogenic biomineral: Shape development and structures of calcium oxalate dehydrate in the presence of polyacrylic acid." *Chemistry – A European Journal* 18: 4000-4009.

Tuason, M. M. and Arocena, J. M. 2009. "Calcium oxalate biomineralization by Piloderma fallax in response to various levels of calcium and phosphorus." *Appl Environ Microbiol.* 75(22): 7079-7085.

Vadlapudi, V., Borah, N., Yellusani, K. R., Gade, S., Reddy, P., Rajamanikyam, M., Vempati, L. N. S., Satya, P. G., Chopra, P.,

Upadhyayula, S. M. and Amanchy, R. 2017. "Aspergillus Secondary Metabolite Database, a resource to understand the Secondary metabolome of Aspergillus genus." *Sci Rep*. 7: 7325.

Verrecchia, E. P., Braissant, O. and Cailleau, G. 2006. "The oxalate-carbonate pathway in soil carbon storage: The role of fungi and oxalotrophic bacteria." In *Fungi in Biogeochemical Cycles* edited by G. M. Gadd. Cambridge: Cambridge University Press.

Walaszczyk, E., Podgórski, W., Janczar-Smuga, M. and Dymarska, E. 2018. "Effect of medium pH on chemical selectivity of oxalic acid biosynthesis by *Aspergillus niger* W78C in submerged batch cultures with sucrose as a carbon source." *Chemical Papers* 72:1089–1093

Wang, H., Sun, K. and Li, A. 2011. "Size-controlled synthesis and characterization of fluorapatite nanocrystals in the presence of gelatin." *J Powder Technol* 209:9–1.

Warscheid, Th. and Braams, J. 2000. "Biodeterioration of stone: a rewiew." *International Biodeterioration and Biodegradation* 46: 343-368.

Wei, Z., Hillier, S. and Gadd, M. G. 2012. "Biotransformation of manganese oxides by fungi: solubilization and production of manganese oxalate biominerals." *Environmental Microbiology* 14 (7): 1744–1752.

Wei, Z., Liang, X., Pendlowski, H., Hillier, S., Suntornvongsagul, K., Sihanonth, P. and Gadd, M. G. 2013. "Fungal biotransformation of zinc silicate and sulfide mineral ores.*" Environmental Microbiology* 5 (8): 2173–2186.

Williams, L. E. and Pittman, J. K. 2010. "Dissecting pathways involved in manganese homeostasis and stress in higher plant cells." In *Cell Biology of Metals and Nutrients. Plant Cell Monographs* edited by Hell, R., Mendel, R. R., 95–117. Berlin: Springer-Verlag.

Wilson, M. J. and Jones, D. 1984. "The occurrence and significance of manganese oxalate in Pertusaria coralline (Lichens)." *Pedobiologia* 26: 373-279.

Wilson, M. J., Jones, D. and McHardy, W. J. 1981. "The weathering of serpentinite by Lecanora atra." *Lichenologist* 13(2): 167-176.

Wilson, M. J., Jones, D. and Russell, J. D. 1980. "Glushinskite, a naturally occurring magnesium oxalate." *Mineralogical magazine* 43: 837-840.

Woronick, C. L. and Johnson, M. J. 1960. "Carbon dioxide fixation by cell-free extracts of *Aspergillus niger*." *J. Biol. Chem.* 235: 9–15.

Wu, Y., Shao, X., Jiao, H., Song, X., He, K. and Li, Z. 2020. "Tracking the fungus-assisted biocorrosion of lead metal by Raman imaging and scanning electron microscopy technique." *J Raman Spectrosc* 1–6.

Xu, B. D. 1989. "The influence of type and concentration of the carbon source on production of citric acid by *Aspergillus niger*." *Appl Microbiol Biotechnol*. 3: 553–558.

Yang, Y., Song, W., Ferrier, J., Liu, F., Csetenyi, L. and Gadd, M. G. 2020. "Biorecovery of cobalt and nickel using biomass-free culture supernatants from *Aspergillus niger*." *Applied Microbiology and Biotechnology* 104: 417-425.

Yang, Y., Ferrier, J., Csetenyi, L. and Gadd, M. G. 2019. "Direct and indirect bioleaching of cobalt from low grade laterite and pyritic ores by *Aspergillus niger*." *Geomycrobiology Journal* 36(10): 940-949.

Zelenskaya, M. S., Rusakov, A. V., Frank-Kamenetskaya, O. V., Vlasov, D. Y., Izatulina, A. R. and Kuz'mina, M. A. 2020. "Crystallization of Calcium Oxalate Hydrates by Interaction of Apatites and Fossilized Tooth Tissue with Fungus *Aspergillus niger*." In *Processes and Phenomena on the Boundary between Biogenic and Abiogenic Nature* edited by Frank-Kamenetskaya, O., Vlasov, D., Panova, E., Lessovaia, S. 581–603. Springer: Cham, Switzerland.

Zhang, A. 2002a. "Citric acid fermentation and heavy metal ions: II. The action of elevated manganese ion concentrations." *Acta Biotechnol*. 22: 375–382.

Zhang, A. 2002b. "Effects of varied phosphorus concentrations on citric acid fermentation by *Aspergillus niger*." *Acta Biotechnol*. 22:383–390.

Zhang, Z. Y. 2007. "Production of lactic acid from renewable materials by Rhizopus fungi." *Biochemical Engineering Journal*. 35: 251–263.

Zhu, T. and Dittrich, M. 2016. "Carbonate Precipitation through Microbial Activities in Natural environment, and Their Potential in

Biotechnology: A Review." *Frontiers in Bioengineering and Biotechnology*. 4: 4.
Zemlyanukhin A. A. 1995. *Metabolism of Plant Organic Acids*. Voronezh: Voronezh University Press. (In Russian).

BIOGRAPHICAL SKETCHES

Dmitry Yu. Vlasov

Affiliation: Saint Petersburg State University

Education:
Diploma, 1981, Saint Petersburg State University, Russia/Department of Biology and Soil Biology
Ph.D., 1987, Plant Protection, All-Union Research Plant Protection Institute/Russia
Dc.Sc., 2008, Mycology, Botanical Institute of Russian Academy of Science/Russia

Business Address: 199034, Russia, Saint Petersburg, Universitetskaya nab. 7/9

Research and Professional Experience:

Ecology, Adaptation and evolution of microscopical fungi (micromycetes), microbial communities in the extreme environment, microbial deterioration and degradation of stone, geomicrobiology biodeterioration of natural and anthropogenic substrates, mechanisms of biodeterioration, ecological aspects of biodeterioration, ecology of fungi, polar research.

Professional Appointments:

1. All-Russian Plant Protection Institute, Lab. Mycology, (1987-1991):

- Scientist
- Leader of the scientific group on soil fungi investigation.

2. Biological Research Institute of Saint Petersburg State University:
   - Chief of the Lab. Mycology (1991-1999).
   - Vice-director on scientific work (1999-2007)
3. Saint Petersburg State University, Biology and Soil Faculty:
   - Head of rhe Dept. of Scientific work of the of Biology and Soil Faculty (2007-2009)
4. Saint Petersburg State University, Biology and Soil Faculty, Dept. of Botany:
   - Professor (the present time)
5. Botanical Institute of Russian Academy of Science:
   - Leading Researcher (the present time)
6. Participating of 49, 52, 54, 56 Russian Antarctic Expedition (2004, 2006, 2008, 2010)
7. President of the St. Petersburg Society of Naturalists

Honors: "Leonard Eyler Medal" (European Academy of Natural Sciences), 2014

Publications from the Last 3 Years:

1. Olga V. Frank-Kamenetskaya, Dmitry Yu. Vlasov, Elena G. Panova, Sofia N. Lessovaia (Editors) *Processes and Phenomena on the Boundary between Biogenic and Abiogenic Nature.* Springer Nature Switzerland AG 2020, 909 p.
2. Olga V. Frank-Kamenetskaya, Dmitry Yu. Vlasov, Vera V. Rytikova (Editors) *The Effect of the Environment on Saint Petersburg's Cultural Heritage Results of Monitoring the Historical Necropolis Monuments.* Springer Nature Switzerland AG 2019 188 p.
3. Olga V. Frank-Kamenetskaya, Gregory Yu. Ivanyuk, Marina S. Zelenskaya, Alina R. Izatulina, Andrey O. Kalashnikov, Dmitry Yu. Vlasov and Evgeniya I. Polyanskaya Calcium Oxalates in

Lichens on Surface of Apatite-Nepheline Ore (Kola Peninsula, Russia)//*Minerals* 2019, 9, 656.

4. Kirtsideli I. Yu., Vlasov D. Yu., Abakumov E. V., Barantsevich E. P., Novozhilov Ju. K., Krylenkov V. A., Sokolov V. T. Airborne fungi in arctic settlement Tiksi (Russian Arctic, coast of the Laptev Sea)//*Czech Polar Reports*. 2017. 7(2).
5. Kirtsideli I. Yu., Vlasov D. Yu., Novozhilov Yu. K., Abakumov E. V., Barantsevich E. P. Assessment of Anthropogenic Influence on Antarctic Mycobiota in Areas of Russian Polar Stations// *Contemporary Problems of Ecology*. 2018. Vol. 11, N 5. P. 449–457.
6. Kuz'mina M. A., Rusakov A. V., Frank-Kamenetskaya O. V., Vlasov D. Y. The Influence of Inorganic and Organic Components of Biofilms with Microscopic Fungi on the Phase Composition and Morphology of Crystallizing Calcium Oxalates//*Crystallography Reports*. 2019. Vol. 64. N1. P. 161-167.
7. Kondratenko Y. A., Vlasov D. Y., Buslaev G. S., Kochina T. A., Shilova O.A. Improving the Bioresistance of Silica-Organic Coatings by Introducing Soft Biocides Based on Intracomplex Compounds of Triethanolamine//*Glass Physics and Chemistry*. Vol. 45, N 5, P. 372-378.
8. Biogenic-Abiogenic Interactions in *Natural and Anthropogenic Systems*. Switzerland, Springer International Publishing, 2016. 543p. (Eds.: O. V. Frank-Kamenetskaya, E. G., Panova, D. Yu. Vlasov).
9. Kirtsideli I. Yu., Vlasov D. Yu. Geographic Distribution and Biodiversity of Microfungi in Soils of the Arctic Regions (at Taimyr Peninsula and Nearby Islands as a Simulation Model of Arctic) In: *Terrestrial Biomes: Geographic Distribution, Biodiversity and Environmental Threats*. Nova Science Publishers 2016:5-25.
10. Sazanova K. V., Vlasov D. Yu., Osmolovskaya N. G., Schiparev S. M., Rusakov A. V. Significance and Regulation of Acids Production by Rock-Inhabited Fungi. In: *Biogenic-Abiogenic*

Interactions in Natural and Anthropogenic Systems (Eds.: O. V. Frank-Kamenetskaya, E. G., Panova, D. Yu. Vlasov). Switzerland, Springer International Publishing, 2016. P. 379-392.

11. Grishkin V. M., Shigorets S. B., Vlasov D. Yu., Miklashevich E. A., Zhabko A. P., Kovshov A. M., Vlasov A. D.. Development of Stone Monuments Monitoring System Using Computer Technology//In: *Biogenic-Abiogenic Interactions in Natural and Anthropogenic Systems* (Eds.: O. V. Frank-Kamenetskaya, E. G., Panova, D. Yu. Vlasov). Switzerland, Springer International Publishing, 2016. P. 415-424.
12. Panova E. G., Vlasov D. Yu., Luodes H., Vlasov A. D., Popova T. A., Zelenskaya M. S.. Granite weathering in urban environment// In: *Biogenic-Abiogenic Interactions in Natural and Anthropogenic Systems* (Eds.: O. V. Frank-Kamenetskaya, E. G., Panova, D. Yu. Vlasov). Switzerland, Springer International Publishing, 2016. P. 345-356.
13. Rusakov A. V., Vlasov A. D., Zelenskaya M. S., Frank-Kamenetskaya O. V., Vlasov D. Ju. The Crystallization of Calcium Oxalate Hydrates Formed by Interaction between Microorganisms and Minerals//In: *Biogenic-Abiogenic Interactions in Natural and Anthropogenic Systems* (Eds.: O. V. Frank-Kamenetskaya, E. G., Panova, D. Yu. Vlasov). Switzerland, Springer International Publishing, 2016. P. 357-378.
14. Abakumov E. V., Parnikoza I. Yu., Vlasov D. Yu., Lupachev A. V. Biogenic-Abiogenic Interactions in Antarctic Ornithogenic Soils//In: *Biogenic-Abiogenic Interactions in Natural and Anthropogenic Systems* (Eds.: O. V. Frank-Kamenetskaya, E. G., Panova, D. Yu. Vlasov). Switzerland, Springer International Publishing, 2016. P. 237-248.
15. Vladimir P. Chelibanov, Aleksandr M. Marugin, Katerina V. Sazanova, Evgeny V. Abakumov, Dmitry Yu. Vlasov, Vera V. Manurtdinova, and Olga V. Frank-Kamenetskaya. Outdoor Environment of the Monuments in the Necropoleis. In: *The Effect of the Environment on Saint Petersburg's Cultural Heritage.*

Results of Monitoring the Historical Necropolis Monuments. Eds: Frank-Kamenetskaya O. V., Vlasov D. Yu., Rytikova V. V. 2019, Springer. P.45-74.

16. Dmitry Yu. Vlasov, Olga V. Frank-Kamenetskaya, Vera V. Manurtdinova, and Marina S. Zelenskaya. Decay of the Monuments. In: *The Effect of the Environment on Saint Petersburg's Cultural Heritage. Results of Monitoring the Historical Necropolis Monuments*. Eds: Frank-Kamenetskaya O. V., Vlasov D. Yu., Rytikova V. V. 2019, Springer. P.75-96.
17. Vera V. Rytikova, Marina S. Zelenskaya, Vera V. Manurtdinova, Iuliia A. Loginova, Khristina V. Shumilova, Olga V. Shchedrova, Ekaterina I. Makeeva, Vadim A. Parfenov, Oleg S. Vereshchagin, Katerina V. Sazanova, Alexey D. Vlasov, Olga V. Frank-Kamenetskaya, Dmitry Yu. Vlasov, and Yuriy M. Pirjutko. Changes in the Necropoleis Monuments' Status Over Time. In: *The Effect of the Environment on Saint Petersburg's Cultural Heritage. Results of Monitoring the Historical Necropolis Monuments*. Eds: Frank-Kamenetskaya O. V., Vlasov D. Yu., Rytikova V. V. 2019, Springer. P.97-144.
18. Olga V. Frank-Kamenetskaya, Dmitry Yu. Vlasov, Vera V. Manurtdinova, Marina S. Zelenskaya, Vadim A. Parfenov, Vera V. Rytikova, Alexey D. Vlasov and Valery M. Grishkin. Monitoring of the State of Stone and Bronze Monuments. In: *The Effect of the Environment on Saint Petersburg's Cultural Heritage. Results of Monitoring the Historical Necropolis Monuments.* Eds: Frank-Kamenetskaya O. V., Vlasov D. Yu., Rytikova V. V. 2019, Springer. P.145-160.
19. Dmitry Yu. Vlasov, Vadim A. Parfenov, Marina S. Zelenskaya, Yuliya V. Plotkina, Valeria M. Geludova, Olga V. Frank-Kamenetskaya and Alexandr M. Marugin. Methods of Monument Protection from Damage and Their Performance. In: *The Effect of the Environment on Saint Petersburg's Cultural Heritage. Results of Monitoring the Historical Necropolis Monuments.* Eds: Frank-

Kamenetskaya O. V., Vlasov D. Yu., Rytikova V. V. 2019, Springer. P.161-178.
20. O. Frank-Kamenetskaya, D. Vlasov, V. Rytikova, V. Parfenov, V. Manurtdinova, M. Zelenskaya. Monitoring of the State of St. Petersburg Stone Monuments and the Strategy of Their Preservation//*Proc. 14t^h International Congress for Applied Mineralogy* (ICAM2019). Springer, 2019. P. 479-483.

Olga V. Frank-Kamenetskaya

Affiliation: Saint Petersburg State University

Education:

Diploma, 1968, Leningrad (St. Petersburg) State University, Russia/Department of crystallography

Ph.D., 1973, Geology and Mineralogy, Leningrad (St. Petersburg) State University/Russia

Dc. Sc., 1998, Geology and Mineralogy, St. Petersburg State University/Russia

Business Address: Universitetskaya nab. 7/9, Saint Petersburg 199034, Russia

Research and Professional Experience: biomineralization, crystal chemistry of minerals, advanced materials, preservation of cultural heritage

Professional Appointments:

Professor of Crystallography department, Institute of Earths Science, Saint Petersburg State University, since 1998;

Vice-president of the St. Petersburg Society of Naturalists;

Member of the Russian Mineralogical Society;

Deputy Chairman of the Organizing Committee of the International Symposium "Biogenic-abiogenic interactions in natural and anthropogenic systems";

Member of the organizing committee of the international conference on crystal chemistry, medical imaging and spectroscopy of minerals and Russian conference with international participation on organic mineralogy.

Honors:

Diploma of the Russian State Committee on Higher Education (1997);

Scholarship of the President and the Government of the Russian Federation for the outstanding scientist (2000-2003);

Winner of the contest "Grant of St. Petersburg" (Soros Professor, 2003);

Winner of the contest RosGeo and Ros bowels of the earth in the field of science and technology innovation in the geological study of Russia (2009);

Medal of St. Petersburg University (2015).

Publications from the Last 3 Years:

1. Olga V. Frank-Kamenetskaya, Dmitry Yu. Vlasov, Elena G. Panova, Sofia N. Lessovaia (Editors) *Processes and Phenomena on the Boundary between Biogenic and Abiogenic Nature*. Springer Nature Switzerland AG 2020, 909 p.
2. Olga V. Frank-Kamenetskaya, Dmitry Yu. Vlasov, Vera V. Rytikova (Editors) *The Effect of the Environment on Saint Petersburg's Cultural Heritage Results of Monitoring the Historical Necropolis Monuments*. Springer Nature Switzerland AG 2019 188 p.
3. Olga V. Frank-Kamenetskaya, Gregory Yu. Ivanyuk, Marina S. Zelenskaya, Alina R. Izatulina, Andrey O. Kalashnikov, Dmitry Yu. Vlasov and Evgeniya I. Polyanskaya Calcium Oxalates in Lichens on Surface of Apatite-Nepheline Ore (Kola Peninsula, Russia)//*Minerals* 2019, 9, 656.

4. Yuliya Bronzova, Miriam Babushkina, Olga Frank-Kamenetskaya, Oleg Vereshchagin, Ira Rozhdestvenskaya, Anatoly Zolotarev Short-range order in Li–Al tourmalines: IR spectroscopy, X-ray single crystal diffraction analysis and a bond valence theory approach//*Physics and Chemistry of Minerals* (2019) 46:815–825.
5. George Kuranov, Anton Nikolaev, Olga Frank-Kamenetskaya, · Nicolay Gulyaev, Olga Volina Physicochemical characterization of human cardiovascular deposits *JBIC Journal of Biological Inorganic Chemistry* (2019) 24:1047-1055.
6. Alina R. Izatulina, Anton M. Nikolaev, Mariya A. Kuz'mina, Olga V. Frank-Kamenetskaya and Vladimir V. Malyshev Bacterial Effect on the Crystallization of Mineral Phases in a Solution Simulating Human Urine//*Crystals* 2019, 9, 259.
7. Kuz'mina, M. A., Rusakov, A. V., Frank-Kamenetskaya, O. V., Vlasov D. Yu. The Influence of Inorganic and Organic Components of Biofilms with Microscopic Fungi on the Phase Composition and Morphology of Crystallizing Calcium Oxalates//*Crystallogr. Rep*. (2019) 64: 161.
8. Aleksei V. Rusakov, Mariya A. Kuzmina, Alina R. Izatulina and Olga V. Frank-Kamenetskaya Synthesis and Characterization of $(Ca,Sr)[C_2O_4]$ nH_2O Solid Solutions: Variations of Phase Composition, Crystal Morphologies and in Ionic Substitutions// *Crystals* 2019, 9, 654.
9. I. E., Kolesnikov, A. M. Nikolaev, E. Lähderanta, O. V. Frank-Kamenetskaya M.A. Kuz'minaa Structural and luminescence properties of Ce3+-doped hydroxyapatite nanocrystalline powders *Optical Materials* 109550 https://doi.org/10.1016/j.optmat.2019.109550.
10. Oleg S. Vereshchagin, Olga V. Frank-Kamenetskaya, Khristina V. Shumilova, Natalia Yu. Khadeeva, Carbonate sediments on decorative fountains in Peterhof, *Russia Environmental Earth Sciences* (2018) 77:56.
11. Khamova, T. V., O. V. Frank-Kamenetskayaa, O. A. Shilovaa, V. P. Chelibanov, A. M. Marugin, E. A. Yasenko, M. A. Kuz'mina,

A. E. Baranchikov, and V. K. Ivanove, *Crystallography Reports*. 2018, Vol. 63, No. 2, pp. 254–260.

12. Alina R. Izatulina, Vladislav V. Gurzhiy, Maria G. Krzhizhanovskaya, Mariya A. Kuz'mina,†atteo Leoni, and Olga V. Frank-Kamenetskaya Hydrated Calcium Oxalates: Crystal Structures, Thermal Stability, and Phase Evolution//*Cryst. Growth Des*. 2018, 18, 5465–5478.
13. Oleg S. Vereshchagin, Olga V. V. Frank-Kamenetskaya, Ira V. Rozhdestvenskaya and Anatoly A. Zolotarev. Incorporation of 3d elements in tourmalines: structural adjustments and stability//*Eur. J. Mineral.* 2018, 30, 917–928.
14. Nikolaev, A., Kolesnikov, I., Frank-Kamenetskaya, O. & Kuz'mina, M. Europium concentration effect on characteristics and luminescent properties of hydroxyapatite nanocrystalline powders//*Journal of Molecular Structure* (2017), 1149, p. 323-331.

Marina S. Zelenskaya

Affiliation: Saint Petersburg State University

Education:

Diploma, 1996, Saint Petersburg State University, Russia/ Faculty of Biology and Soil Biology

Ph.D., 2000, Mycology, Botanical Institute of Russian Academy of Science/Russia

Business Address: Saint Petersburg, Russia, 199034, Universitetskaya nab. 7/9

Research and Professional Experience:

Ecology, Adaptation and evolution of microscopical fungi (micromycetes), microbial biodeterioration of stone, geomycology, mechanisms of biodeterioration of natural and anthropogenic substrates, preservation of cultural heritage, biomineralization

Professional Appointments:

1. Biological Research Institute of Saint Petersburg State University (1998-2015) Researcher of the Lab. Mycology
2. Saint Petersburg State University, Biology Faculty, Department of Botany, Senior Researcher (2015 - the present time)

Publications from the Last 3 Years:

1. Vlasov Dmitry Yu, Panova Elena G, Zelenskaya Marina S, Vlasov Alexey D, Sazanova Katerina V, Rodina Oksana A, Pavlova Olga A Biofilms on granite rapakivi in natural outcrops and urban environment: Biodiversity, metabolism and interaction with substrate.//*Lecture Notes in Earth System Sciences*. Springer, 2020. p. 535-559 (Lecture Notes in Earth System Sciences). ISBN 978-3-030-21613-9 ISBN 978-3-030-21614-6 (eBook) https://doi.org/10.1007/978-3-030-21614-.
2. Zelenskaya Marina S, Rusakov Aleksei V, Frank-Kamenetskaya Olga V, Vlasov Dmitry Yu, Izatulina Alina R, Kuz'mina Mariya A Crystallization of calcium oxalate hydrates by interaction of apatites and fossilized tooth tissue with fungus *Aspergillus niger*.//*Lecture Notes in Earth System Sciences*. Springer, 2020. p. 581-603 (Lecture Notes in Earth System Sciences). ISBN 978-3-030-21613-9 ISBN 978-3-030-21614-6 (eBook) https://doi.org/10.1007/978-3-030-21614-.
3. Zaitsev Anatoly N, Vlasov Dmitry Yu, Zelenskaya Marina S, Zaitseva Olga A, Pavlova Olga A, Chakhmouradian Anton R, Savchenok Anton I, Leach Larissa, Leach Michael, Mwankunda Joshua Microbiology of the laetolil tuff 7 with 3.66 Ma australopithecus afarensis footprints, ngorongoro conservation area, Tanzania//*Lecture Notes in Earth System Sciences*. Springer, 2020. p. 669-692 (Lecture Notes in Earth System Sciences). ISBN

978-3-030-21613-9 ISBN 978-3-030-21614-6 (eBook) https://doi.org/10.1007/978-3-030-21614-.

4. Shilova Olga, Vlasov Dmitry Yu, Zelenskaya Marina S, Ryabusheva Yuliya V, Khamova Tamara V, Glebova Irina B, Sinelnikov Alexandr A, Marugin Alexandr M, Frank-Kamenetskaya Olga V Sol-gel derived Tio_2 and epoxy-titanate protective coatings: Structure, property, fungicidal activity and biomineralization effects.//*Lecture Notes in Earth System Sciences*. Springer, 2020. p. 619-638 (Lecture Notes in Earth System Sciences). ISBN 978-3-030-21613-9 ISBN 978-3-030-21614-6 (eBook) https://doi.org/10.1007/978-3-030-21614-.
5. Frank-Kamenetskaya O. V., Vlasov D. Yu., Manurtdinova V. V., Zelenskaya M. S., Parfenov V. A., Rytikova V. V., Vlasov A. D., Grishkin V. M. Monitoring of the state of stone and bronze monuments//*The Effect of the Environment on Saint Petersburgs Cultural Heritage. Results of Monitoring the Historical Necropolis Monuments*. Сер. "Geoheritage, Geoparks and Geotourism' Cham, 2019. P. 145-159. ISBN 978-3-319-79071-8, ISBN 978-3-319-79072-5 (eBook) https://doi.org/10.1007/978-3-319-79072-5.
6. Vlasov D. Yu., Parfenov V. A., Zelenskaya M. S., Plotkina Yu. V., Geludova V. M., Frank-Kamenetskaya O. V., Marugin A. M. Methods of monument protection from damage and their performance//*The Effect of the Environment on Saint Petersburgs Cultural Heritage. Results of Monitoring the Historical Necropolis Monuments.* Сер. "Geoheritage, Geoparks and Geotourism" Cham, 2019. P. 161-178. ISBN 978-3-319-79071-8, ISBN 978-3-319-79072-5 (eBook) https://doi.org/10.1007/978-3-319-79072-5.
7. Vlasov D. Yu., Frank-Kamenetskaya O. V., Manurtdinova V. V., Zelenskaya M.S. Decay of the monuments//*The Effect of the Environment on Saint Petersburgs Cultural Heritage. Results of Monitoring the Historical Necropolis Monuments.* Сер. "Geoheritage, Geoparks and Geotourism" Cham, 2019. P. 75-90. ISBN 978-3-319-79071-8, ISBN 978-3-319-79072-5 (eBook) https://doi.org/10.1007/978-3-319-79072-5.

8. Rytikova V. V., Zelenskaya M. S., Manurtdinova V. V., Loginova Iu. A., Shumilova Kh. V., Shchedrova O. V., Makeeva E. I., Parfenov V. A., Vereshchagin O. S., Sazanova K. V., Vlasov A. D., Frank-Kamenetskaya O. V., Vlasov D. Yu., Pirjutko Yu. M. Changes in the necropoleis monuments status over time//*The Effect of the Environment on Saint Petersburgs Cultural Heritage. Results of Monitoring the Historical Necropolis Monuments*. Сер. "Geoheritage, Geoparks and Geotourism" Cham, 2019. P. 97-143. ISBN 978-3-319-79071-8, ISBN 978-3-319-79072-5 (eBook) https://doi.org/10.1007/978-3-319-79072-5.
9. Olga V. Frank-Kamenetskaya, Gregory Yu. Ivanyuk, Marina S. Zelenskaya, Alina R. Izatulina, Andrey O. Kalashnikov, Dmitry Yu. Vlasov and Evgeniya I. Polyanskaya Calcium Oxalates in Lichens on Surface of Apatite-Nepheline Ore (Kola Peninsula, Russia)//*Minerals* 2019, 9, 656; doi:10.3390/min9110656.
10. Vlasov Dmitry Yu., Teshebaev Sh. B., Zelenskaya M. S., Kirtsideli I. Yu., Ryabusheva Yu. V. Mycological Damage of Materials in the Indoor Environment as a Risk Factor for the Health of Polar Explorers//*Gigiena i sanitariia* 2019, 98(1):P. 17-21 DOI:10.18821/0016-9900-2019-98-1-17-21.
11. Ianina Bogdanova (Pagoda), Anatoly Pautov, Marina Zelenskaya, Elena Krylova, and Dmitry Vlasov Seasonal changes in the phylloplane of genus Gnetum (Gnetaceae) representatives in greenhouse conditions//*Biological Communications* 2020, 64(4), p. 260-269 doi:10.21638/spbu03.2019.405.
12. Kirtsideli I. Yu., Vlasov D. Yu., Zelenskaya M. S., Barantsevich E. P., Novozhilov Yu. K., Krylenkov V. A., Churkina I. V., Sokolov V. T. Anthropogenic modification of mycobiota on Vize island (in the Kara Sea)//*Gigiena i Sanitariya* 2018, 97(11), p. 1058-1063.
13. Kirtsideli I. Yu., Teshebaev Sh. B., Vlasov D. Yu., Novozhilov Yu. K., Abakumov E. V., Barantsevich E. P., Krylenkov V. A., Zelenskaya M. C. Changes in microbial communities in primary soil and ground under the anthropogenic influence on the territory

around antarctic station "Mirny"//*Gigiena i Sanitariya* 2017, 96(10), c. 949-955.

14. Zelenskaya M. S., Sidelnikova M. V., Panova E. G., Pautov A. A., Krylova E. G., PagodaYa. O., Vlasov D. Yu. Phylloplane Fungi In The Urban Environmentms//*BIOSFERA* 2017, 9 (2), p. 136-151 doi: 10.24855/BIOSFERA.V9I2. 35315137.

Katerina V. Sazanova

Affiliation: Komarov Botanical Institute RAS, The Archive of the Russian Academy of Sciences,

Education:

Diploma, 2011, Saint Petersburg State University, Russia/Department of Biology and Soil Biology

Ph.D., 2015, Mycology, Komarov Botanical Institute RAS/Russia

Business Address: Universitetskaya nab. 7/9, Saint Petersburg 199034, Russia

Research and Professional Experience: Biomineralization, biochemistry of fungi, metabolomics, organic acids, preservation of cultural heritage

Professional Appointments:

Researcher at the Laboratory of Analytical Phytochemistry in Komarov Botanical Institute RAS (2008 - the present time).

Publications from the Last 3 Years:

1. Sazanova K. V., Prutzeva N. V., Shavarda A. L. Cultural and metabolomic studies of a new phtalides producer, *Lignomyces vetlinianus* (*Agaricomycetes*)//*International Journal of Medicinal Mushrooms.* 2018. T. 20. № 11. C. 1031-1045. doi: 10.1615/ IntJMedMushrooms.2018028687.

2. Sazanova K. V., Senik S. V., Kirtsideli I. Yu., and A. L. Shavarda. Metabolomic Profiling and Lipid Composition of Arctic and Antarctic Strains of Micromycetes Geomyces pannorum and Thelebolus microsporus Grown at Different Temperatures//*Microbiology (Mikrobiologiya).* 2019. T. 88. № 3. C. 1-12. doi: 10.1134/S002636561903011X.
3. Chelibanov V. P., Marugin A. M., Sazanova K. V., Abakumov E. V., Vlasov D. Yu., Manurtdinova V. V., Frank-Kamenetskaya O. V. Outdoor environment of the monuments in the Necropoleis. In: The Effect of the Environment on Saint Petersburg's Cultural Heritage Results of Monitoring the Historical Necropolis Monuments. "*Geoheritage, Geoparks and Geotourism*" *Cham,* 2019. C. 45-74.
4. Rytikova V. V., Zelenskaya M. S., Manurtdinova V. V., Loginova Iu. A., Shumilova Kh. V., Shchedrova O. V., Makeeva E. I., Parfenov V. A., Vereshchagin O. S., Sazanova K. V., Vlasov A. D., Frank-Kamenetskaya O. V., Vlasov D. Yu., Pirjutko Yu. M. Changes in the necropoleis monuments' status over time In: The Effect of the Environment on Saint Petersburg's Cultural Heritage Results of Monitoring the Historical Necropolis Monuments. "*"Geoheritage, Geoparks and Geotourism" Cham*, 2019.C.97-143.
5. Sazanova K. V., Vlasov D. Yu., Shavarda A. L., Zelenskaya M. S. Biochemical composition of rock-inhabited communities in anthropogenic and natural areas In: *VI International Symposium "Biogenic - abiogenic interactions in natural and anthropogenic systems"* devoted to the 150th anniversary of the Saint-Petersburg Naturalists Society. 2018. C. 161-162.

Aleksei V. Rusakov

Affiliation: Saint Petersburg State University

Education: Diploma, 2012, Saint Petersburg State University, Russia/Institute of Earth Sciences

Business Address: Universitetskaya nab. 7/9, Saint Petersburg 199034, Russia

Research and Professional Experience: Biomineralization, biomimetics, preservation of cultural heritage, science visualization

Professional Appointments: research engineer of Crystallography dep., Institute of Earths Science, Saint Petersburg State University (2013-2017, 2019-present)

Publications from the Last 3 Years:

1. Zelenskaya, M. S.; Rusakov, A. V.; Frank-Kamenetskaya, O. V.; Vlasov, D. Y.; Izatulina, A. R.; Kuz'mina, M. A. Crystallization of Calcium Oxalate Hydrates by Interaction of Apatites and Fossilized Tooth Tissue with Fungus *Aspergillus niger*. In *Processes and Phenomena on the Boundary Between Biogenic and Abiogenic Nature*; Frank-Kamenetskaya, O., Vlasov, D.; Panova, E., Lessovaia, S., Eds.; *Lecture Notes in Earth System Sciences*; Springer: Cham, Switzerland, 2020; pp. 581–603 doi:10.1007/978-3-030-21614-6_31.
2. Kuz'mina, M. A., Rusakov, A. V., Frank-Kamenetskaya, O. V. Vlasov D. Yu. The Influence of Inorganic and Organic Components of Biofilms with Microscopic Fungi on the Phase Composition and Morphology of Crystallizing Calcium Oxalates//*Crystallogr. Rep.* (2019) 64: 161. doi:10.1134/S106377 4519010176.

3. Aleksei V. Rusakov, Mariya A. Kuz'mina, Alina R. Izatulina and Olga V. Frank-Kamenetskaya Synthesis and Characterization of $(Ca,Sr)[C_2O_4]\cdot nH_2O$ Solid Solutions: Variations of Phase Composition, Crystal Morphologies and in Ionic Substitutions//*Crystals* 2019, 9, 654; doi:10.3390/cryst9120654.
4. Sazanova, K. V.; Vlasov, D. Yu.; Osmolovskay, N. G.; Schiparev, S. M.; Rusakov, A. V. Significance and regulation of acids production by rock-inhabited fungi. In *Biogenic—Abiogenic Interactions in Natural and Anthropogenic Systems*; Frank-Kamenetskaya, O. V.; Panova, E. G.,; Vlasov, D.Yu., Eds.; Springer Verlag: Berlin, Germany, 2016. pp. 379-392. doi:10.1007/978-3-319-24987-2_29.
5. Rusakov, A. V.; Vlasov, A. D.; Zelenskaya, M. S.; Frank-Kamenetskaya, O. V.; Vlasov, D. Yu. The crystallization of calcium oxalate hydrates formed by interaction between microorganisms and minerals. In *Biogenic — Abiogenic Interactions in Natural and Anthropogenic Systems*; Frank-Kamenetskaya, O. V.; Panova, E. G., Vlasov, D. Yu. Springer Verlag: Berlin, Germany, 2016. pp. 357-377.

Alina R. Izatulina

Affiliation: Saint Petersburg State University

Education: Diploma, 2006, Omsk State University, Russia/Department of chemistry

Ph.D., 2016, St. Petersburg State University/Russia

Business Address: Universitetskaya nab. 7/9, Saint Petersburg 199034, Russia

Research and Professional Experience: biomineralization, crystal chemistry of minerals, crystal chemistry of organic minerals, oxalate minerals.

Professional Appointments:

Senior Researcher, Crystallography dep., Institute of Earths Science, Saint Petersburg State University

Member of the St. Petersburg Society of Naturalists;

Member of the Russian Mineralogical Society.

Publications from the Last 3 Years:

1. Izatulina A. R., Gurzhiy V. V., Krzhizhanovskaya M. G., Kuz'mina M. A., Leoni M., Frank-Kamenetskaya O. V. Hydrated calcium oxalates: crystal structures, thermal stability and phase evolution//*Crystal Growth and Design*. 2018. V. 18. P. 5465-5478.
2. Olga V. Frank-Kamenetskaya, Gregory Yu. Ivanyuk, Marina S. Zelenskaya, Alina R. Izatulina, Andrey O. Kalashnikov, Dmitry Yu. Vlasov and Evgeniya I. Polyanskaya Calcium Oxalates in Lichens on Surface of Apatite-Nepheline Ore (Kola Peninsula, Russia)//*Minerals* 2019, 9, 656.
3. Alina R. Izatulina, Anton M. Nikolaev, Mariya A. Kuz'mina, Olga V. Frank-Kamenetskaya and Vladimir V. Malyshev Bacterial Effect on the Crystallization of Mineral Phases in a Solution Simulating Human Urine//*Crystals* 2019, 9, 259.
4. Alina R. Izatulina, Vladislav V. Gurzhiy, Maria G. Krzhizhanovskaya, Nikita V. Chukanov and Taras L. Panikorovskii. Thermal Behavior and Phase Transition of Uric Acid and Its Dihydrate Form, the Common Biominerals Uricite and Tinnunculite//*Minerals* 2019, 9, 373.
5. Aleksei V. Rusakov, Mariya A. Kuz'mina, Alina R. Izatulina and Olga V. Frank-Kamenetskaya Synthesis and Characterization of $(Ca,Sr)[C_2O_4]\cdot nH_2O$ Solid Solutions: Variations of Phase Composition, Crystal Morphologies and in Ionic Substitutions//*Crystals* 2019, 9, 654.

6. Gurzhiy, V. V., Tyumentseva, O. S., Izatulina, A. R., Krivovichev, S. V., Tananaev, I. G. Chemically Induced Polytypic Phase Transitions in the $Mg[(UO_2)(TO_4)2(H_2O)](H_2O)_4$ (T = S, Se) System//*Inorganic Chemistry* Volume 58, Issue 21, 2019, Pages 14760-14768.
7. Zelenskaya, M. S.; Rusakov, A. V.; Frank-Kamenetskaya, O. V.; Vlasov, D. Y.; Izatulina, A. R.; Kuz'mina, M. A. Crystallization of Calcium Oxalate Hydrates by Interaction of Apatites and Fossilized Tooth Tissue with Fungus *Aspergillus niger.* In Processes and Phenomena on the Boundary between Biogenic and Abiogenic Nature; Frank-Kamenetskaya, O., Vlasov, D.; Panova, E., Lessovaia, S., Eds.; *Lecture Notes in Earth System Sciences*; Springer: Cham, Switzerland, 2020; pp. 581–603.

In: *Aspergillus niger*
Editor: Eva Baughan
ISBN: 978-1-53618-080-0

Chapter 2

USE AND APPLICATIONS OF *ASPERGILLUS NIGER* FOR THE DEVELOPMENT OF ENZYMES AND PRODUCTS OF BIOTECHNOLOGICAL INTEREST

Luciane Maria Colla[1], Naiara Elisa Kreling[1], Mateus Torres Nazari[1], Paola Manfredini[1], Thaís Strieder Machado[1], Jorge Alberto Vieira Costa[1], Thamarys Scapini[2], Aline Frumi Camargo[2], Karina Paula Preczeski[2], Fábio Spitza Stefanski[2], Bianca D'arck Melo Cavalcante[3], Gislaine Fongaro[3] and Helen Treichel[2,*]

[1]University of Passo Fundo, Passo Fundo, RS, Brazil
[2]Federal University of Fronteira Sul, Erechim, RS, Brazil
[3]Federal University of Santa Catarina, Florianópolis, SC, Brazil

[*] Corresponding Author's Emai: helentreichel@gmail.com.

Abstract

Filamentous fungi of the genus Aspergillus are potential sources of several enzymes and bioproducts. Aspergillus encompasses several distinct species, many of which possess promising properties and mechanisms for various applications. They are found naturally in diverse environments, exhibiting stability, selectivity, and easy assimilation of alternative carbon sources, making this fungus valuable for industrial purposes. *Aspergillus niger* has many properties that are industrially useful, including the ability to naturally accumulate compounds, and to produce biologically active secondary metabolites as well as many proteins and enzymes with biotechnological potential. Among the biotechnological and industrial enzymes produced by *Aspergillus niger* are amylases, cellulases, pectinases, peroxidases, lipases, xylanases, proteases, and laccases, used for pharmaceutical, textiles, food, beverages, and biorefineries. These biocatalysts often replace chemical hydrolysis processes or are used in the production of compounds such as peptides that have antibacterial, antifungal and antioxidant properties. They are also used in the production of products such as biosurfactants and fungal chitosan. This chapter will review recent advances in the prospecting of biocompounds and products with the most relevant biological and biocatalytic activity produced by *Aspergillus niger*. The challenges and future perspectives of biotechnological process extensions using *Aspergillus niger* will also be discussed.

Keywords: enzymes, bioproducts, biopeptides, fungal chitosan

Introduction

The metabolic diversity of the fungi and the wide range of ecological niches they inhabit mean that many species, especially the basidiomycetes, have significant potential as sources of enzymes [1]; *Aspergillus* species produce a variety of enzymes for different areas of interest [2–4].

Solid-state fermentation (SSF) and submerged fermentation (SmF) are the primary methods for producing enzymes [5]. SmF occurs in liquid media containing the necessary nutrients. The advantages of this process include the homogeneity of the environment, the possibility of controlling parameters such as temperature and pH [6], as well as ease of handling and greater control of environmental factors [7]. SSF mimics the natural habitat of filamentous fungi, allowing for production of these biomolecules with greater activity and stability, at lower cost through the use of agroindustrial residues and simple equipment, in addition to being environmentally correct, with less effluent produced [6, 8]. SSF has advantages over SmF, including lower costs and time of cultivation, in addition to higher yields and greater product stability [9].

Regarding cultivation conditions, most fungi show better performance for enzyme production in acidic environments (pH 4.0 to 6.5). *A. niger* grows in the wide temperature range of 6°C to 47°C; however, the ideal growth temperature is 35–37°C [2]. Among the enzymes producing fungi, the genus *Aspergillus* has a wide range of species used for processing: *A. niger, A. oryzae, A. terreus, A. fumigatus, A. foetidus, A. sojae,* and *A. candidus*. Of these, *A. niger, A. oryzae, A. sojae* are generally considered safe by the Food and Drug Administration (FDA) and are approved for use in the food industry [10].

Enzymes and other bioproducts produced by microbial fermentation are relevant for the advancement of biotechnology. *A. niger* has the capacity to carry out a series of transformations; for these reasons, it is the target of much research related to its metabolism and trends [11, 12]. The aim of this chapter is to highlight the use of the fungus *Aspergillus niger* to obtain enzymes amylase, peroxidase, pectinase, keratinase, cellulase, lipase, xylanase, protease, laccase, as well as recent advances in the prospecting of biocompounds and products with the most relevant biological and biocatalytic activity.

ENZYMES FROM *ASPERGILLUS NIGER* AND APPLICATIONS

Amylase

The advancement of biotechnology has enabled the use of microbial amylases in the pharmaceutical, textile, biofuels, food and beverage areas, detergent and paper manufacturing [5, 13]. In particular, amylases from microbial sources have been shown to be safe while presenting higher thermal stability; for these reasons, they are well suited for industrial applications [14, 15]. Among microbial enzymes, filamentous fungi are recognized as a potential sources of amylolytic enzymes, utilizing various residues or byproducts from the agricultural industry [16–18].

Fungi species of the genus *Aspergillus* are the most exploited by the scientific community for amylase production [3].

Of these, *A. niger* demonstrates potential for the production of stable amylases of industrial interest. The development of fungal amylases with the capacity to withstand adverse reactions, as in the presence of acids, could fill some gaps and supplement enzymatic processes with greater activity and cost reduction. In this sense, *A. niger* is a source of acid-stable amylase. These characteristics make the acid-stable amylase a product with high added value for the industry, mainly for the baking industry [19].

Fungal amylases are secreted extracellularly. this is advantageous as it facilitates hydrolysis reactions with substrates, including the most difficult ones, including residues. The first amylase was produced many decades ago; however, *A. niger* has been the most relevant in biotechnology; combined with genetic engineering, new strains may improve medicines and industrial products [20].

Peroxidase

Peroxidases are extracellular glycoproteins that catalyze H_2O_2, produced by lignolytic microorganisms. Because of their non-specific characteristics, they are able to catalyze phenols and non-phenols, including lignin and toxic components; these have applications in industries such as paper and cellulose, textiles, food, bioremediation, cosmetics, and others [21, 22].

A. niger is a great producer of peroxidases that can be produced under various cultivation conditions to be applied in biotechnology [23]. The production of lignin peroxidase and manganese peroxidase can be carried out in the solid state, using H_2O_2 as substrate, obtaining maximum activity in a period of 12 days or even producing them from prickly palm cactus husk, as they are ligninolytic enzymes. They remain active after freezing of crude extracts [24, 25]. They can also be purified using 70% ammonium sulfate and chromatography, reaching 7.75-fold purification [24].

One of the applications of *A. niger* peroxidases is the degradation of toxic compounds such as dyes. Congo red biodegradation and detoxification with a maximum discoloration of 97% in 6 days has already been described. Phytotoxicity and microtoxicity tests revealed that the metabolites generated in the degradation were less toxic than the original dye [26]. Peroxidases combined with laccases mineralized Thiazole Yellow G dye, with 98% discoloration in 5 days [27]. Less common applications of these enzymes involve the use of *A. niger* as an elicitor in plants, aiming to increase the synthesis of bioactive compounds responsible for plants defenses [28] and the use of peroxidases for bioremediation of crude oil [29], which demonstrate a large area of activity of oxidative enzymes, aiming to solve numerous current problems.

Pectinase

Pectinases are enzymes that depolymerize pectic substances, a class of complex glycosidic polysaccharide, via hydrolysis and esterification reactions [30]. They are an important enzymes for the biotechnological sector, with many potential applications, mainly in the food and beverage industry [31]. These enzymes can be used for wastewater treatment in the paper and pulp industry, natural fiber treatment in the textile industry, as well as in new industries such as bioethanol and oligosaccharide production. Pectinases represent almost 25% of the global food enzyme market and can be obtained from pectin-containing wastes such as citrus and orange, apple, grape, banana or other fruit residues without the need for harsh pre-treatments [30, 32].

A. niger has polygalacturonic acid (PG) encoding genes and they express PG isoenzymes that differ according to substrate specificity, cleavage rate, pH and temperature [32]. The fungus enjoys GRAS status

(Generally Regarded as Safe), according to the United States Food and Drug Administration [30, 31].

Pectinases from *A. niger* help to degradation peels, straws, and bagasse of fruits and vegetables used for the production of bioethanol, with a higher yield, converted polysaccharides into reducing sugar [33].

The interaction fungal strains and agro-food industry wastes on pectinases production demonstrated an efficiency; therefore, this process have been optimized and scaled-up [11]. In a study using orange peels as a substrate at a semi-industrial scale and *A. niger* optimized for enzyme production using the Plackett-Burman Experimental Design (PBED) and Box-Behnken (BB), improved pectinase production was observed [30].

The attributes of pectinases are focused mostly on studies with recombinant DNA technology. Pectinolytic genes have been cloned and expressed in various organisms [11], producing highly-pure enzymes that are stable and produce high yields [34]. The production of pectinase with *A. niger* strains has yet many applications worth exploring; nevertheless, it is certain that the use of various agro-industrial residues as substrates for enzymatic production is low-cost, simple and efficient, and should inspire the development of green energy and bioproduct generation.

Keratinase

Keratinases hydrolyze keratinous materials that are formed by a highly rigid structure of α and β-keratin. These microbial enzymes, being low cost and free from contaminants, are used in several applications [35, 36]. *A. niger* is also potentially keratinolytic. This enzyme group has important activity against solid residues; nevertheless, studies of its applications remain superficial, and are restricted to enzymatic production from residues. Keratinases from *A. niger* (ANK) are produced primarily by solid-state fermentation (FES) using various types of keratin, mainly feathers and hair as the only source of nutrients.

The activities in this system can vary substantially, depending on the place from which was *A. niger* obtained and the fermentative substrate used, giving values of keratinolytic activity such as 13 U/g [37], 11.8 U/mL [38] and 172 U/mL [39]. Submerged systems (SmF) are less common for fungi, but with easier production of enzymes, 3 U/mL [40], 39.7 μ/mL [41] and 60 U/mL [39] for example, however, obtaining values of keratinolytic activities lower than those achieved by FES.

The cloning of *A. niger* keratinolytic protease genes to be expressed in *Escherichia coli* has been reported, reaching activity of 21.2 U/mg. These cases occur because the bacteria have accelerated metabolism, reaching high yields in short periods of time when compared to those of fungal production [42], but not dispensing its use in this process.

Cellulase

Cellulases are a set of four enzymatic categories that act synergistically to degrade cellulose, namely endoglucanase (EG), cellobiohydrolase (CBH), β-glucosidase (BG) and lytic polysaccharide mono-oxygenase (LPMO) [43].

These enzymes are currently applied in the production of second-generation ethanol, in which the biofuel is produced from lignocellulosic materials, including bagasse and plant residues [44]. These lignocellulosic materials require a hydrolysis step that can be performed by this enzymatic cocktail in the addition of water, using potable water in current industrial processes. The problem of this water footprint could be mitigated with the total or partial replacement of freshwater with saline water [45]. The use of saline water could harm ethanol yield; however, studies are bringing marine *A. niger* as a producer of cellulases cocktail, tolerant to salt, capable of acting in different saline concentrations. Nevertheless, various different salt concentrations make these enzymes thermostable, improve their catalytic activity and increase the affinity of the enzymes with the substrate [46–48].

Cellulase production is usually carried out using agro-industrial residues, however new residues are being tested with success of enzymatic production utilizing *A. niger*, as is the case of the use of textile residues as a fermentative substrate. This process greatly encourages research and science in the biotechnological area using optimized processes and developing a stronger circular economy model, because the remaining fibers can still be recycled into value-added products [49].

There is also the use of mixed cultures, such as *Trichoderma reesei* and *A. niger* for the production of cellulolytic enzymes to increase the cellulose degradation capacity, considerably increasing levels of ethanol production using pre-treated rice straw and corn straw with sodium hydroxide as enzymatic production substrates [50, 51]. The use of materials rich in lignin hinders the access of enzymes in the cellulose; however, the hydrolysis of the cellulose can be facilitated by reduction of its crystallinity using pre-treatment techniques with NaOH prior to the hydrolytic process by *A. niger* [52]. The fact that delignification facilitates the action of cellulases was also observed by Wang et al. [53].

The immobilization of *A. niger* celluloses is also promising, using various support materials without much loss of activity during enzymatic cycles [54, 55]. An interesting study identified *A. niger* inhabiting solid urban waste, presenting high potential for the production of cellulases in the screening processes, which can be used together with other fungi isolated from the same locations, for the treatment of solid urban waste in the near future [56].

Lipase

Lipases are poorly soluble in water, belonging to the class of hydrolytic enzymes that acting at the water-lipid interface [57, 58]. They act in aqueous and non-aqueous systems, are stable in organic solvents, at various pHs and in extreme temperatures [59]. Produced in liquid or solid media, the characteristics of lipases also depend on the fungal strain used. *A. niger* is an excellent producer of these enzymes and has been studied using oil cakes (oil meal) and fibrous residues as substrates [60–63].

Concentration and enzymatic immobilization techniques can increase the enzymatic activity of several enzymes. Lipase of *A. niger* (ANL) can be concentrated using membranes [64] with the application of organic salts and solvents [65] or by the combined action of reverse micelle extraction and acetone precipitation [66]. The immobilization technique provides for greater stability and enzymatic recovery, essential steps in the industrial implementation of these biocatalysts.

A promising technique for increasing enzyme activity can be achieved by combining the sonication process with subsequent immobilization. In a study, sonication increased the ANL 1.6-fold, causing change and/or activation of the enzymatic conformation; this highly active enzyme was immobilized in zeolite, increasing to 3.2-fold with respect to free lipase [67]. Another ANL was immobilized in macroporous acrylic resin to de-acidify soy sauce residue, achieving reuse of 15 cycles, maintaining 92.08% of its initial activity. It was also possible to desorb immobilized enzymes, adding a new enzyme to the support, significantly reducing the costs, obtaining a new process with enormous potential for environmental/industrial use [68]. Immobilized lipases are achieving substantial activity and stability during several processes, maintaining 83% of activity after ten successive cycles of biodiesel synthesis [69] or producing 47.3% of biodiesel yield, maintaining 57% of the initial activity after four cycles of interesterification [70].

Because of their affinity in the hydrolysis of various oils, ANLs are widely used for biodiesel synthesis, varying fermentative pies in enzymatic production [70], obtaining powder lipases [71] or biodiesel synthesis with whole-cell lipase using microalgae [72]. The scale expansion of ANL production is required for the use of the enzyme extracts in industrial applications. A recent study used a mixture of various substrates, including agricultural residues, for production in a bioreactor containing 1 kg of the substrate, reaching 208.79 ± 6.26 U/gds of lipase activity, showing the ability to include biological processes in the replacement of chemicals [73].

Xylanase

Xylanases belong to the class of glycosidases that release lignin by hydrolysis of lignin-carbohydrate complex, producing xylooligosaccharide and pentose sugars [74, 75]. These enzymes have been widely used in several fields, including pulp biobleaching, biofuel production, clarification of juices and wines, improvement of bread-quality and animal feed processing [76].

A. niger is a xylanase producer using agro-industrial waste as feedstock for the process of production, which has been considered a potential low-cost lignocellulosic substrate [74, 76, 77]. Commercial xylanases of *A. niger* are widely used in pulp bleaching, biofuel production, clarification of juices and wines, improvement of bread-quality, food and animal feed industries, textile industries and in water management [76, 78]. Mainly in the paper and pulp industries, xylanases are valuable because they reduce usage of chemical bleaching agents. Currently, xylanases are applied in the development of products with greater added value, including biomolecule synthesis [79]. For example, β-xylosidases catalyze cleavage of oligosaccharides into xylose; this enzyme can be used for ethanol production or it can be hydrolyzed by xylose reductase into xylitol, an expensive non-caloric sweetener [80].

Some studies with *A. niger* are focusing on strategies to improve xylanase activity using experimental models, enhancing xylanase production by more than 100% [79]. Li et al. [75], searching for better enzyme performance of a recombinant xylanase from *A. niger* nl-1, selected a site-directed mutagenesis method and observed an effect of disulfide in the structure of xylanase. Their results suggest that these amino acid residues are important to the thermostability and pH stability of xylanase from *A. niger* nl-1, which can raise its potential interest for industrial applications.

In addition, techniques of purification have been applied to separation of xylanase from fungal extracts. Precipitation with natural polymers increase the degree of purification of xylanases from *A. niger*, which make this method suitable to be included in a downstream process, providing such advantages as simplicity, scalability, and ability to concentrate the enzyme in a single step process, as well as the use non-toxic reagents [76].

Considering the trend of studies in the field of biomolecules development, the xylanases produced by *A. niger* need to be explored and investigated to a greater extent. The biofabrication of high-value products such as nanoparticles valorize the use of xylanases and agro-waste residues. Recent studies demonstrated the potential use of *A. niger* xylanases for gold nanoparticle production with biological activities that stimulate the development of this product for use as a biomedical agent [79].

Protease

Fungal proteases attract wide attention from biotechnologists because of their growth on low-cost substrates, secretion of large amounts of enzymes in the medium [81], ease of biomass filtering and high growth rates [82]. Via fungal metabolic and physiological mechanisms, proteases may have industrial applications to break down complex molecules and produce new products with high added value [83].

Filamentous fungi, when applied as an initial culture in fermented food, beverage and bioprocess industries, secrete enzymes directly into the medium that benefits from their hydrolytic activity, without the need for additional enzymatic recovery steps, eliminating the inconvenience of their extraction and exogenous application, creating a process that is more economical in terms of time and cost [82].

A. niger is considered one of the most important sources of fungal proteases. Genome sequencing shows that *A. niger* has 198 proteins involved in the proteolytic degradation process. It produces abundant quantities of aspartyl endoproteases and serine carboxypeptidases [84]. Data found in the literature indicate that *Aspergillus* synthesizes acid [85], neutral [86] and alkaline proteases [87].

Proteases from *A. niger* after purification have high thermostability, high solubility, and substantial emulsification capacity, able to catalyze degradation of various commercial substrates [88], and able to degrade complex polymers of casein, a protein present in milk [89].

A. niger secretes enzymes with distinct biochemical properties in response to the substrate used in fermentation, showing considerable tolerance to NaCl, for example, when produced in wheat bran in SSF [90].

Proteases produced by *A. niger* are used for gluten degradation [91]. In the juice and alcohol industry, *Aspergillus* proteases can be used to reduce unwanted turbidity formed from proteins. In beer production, for example, the protease produced by *A. niger* efficiently inhibited the formation of this turbidity without affecting beer foam [40].

The primary trends and challenges in *A. niger* protease production are related to the type of enzyme produced, its biochemical characteristics, its applicability in industries, and large-scale production. Because of the high demand for proteases, these studies must be continuously developed in order to obtain proteolytic enzymes with various biochemical properties, including high activity and stability at alkaline pH, high temperature, presence of surfactants, metals, organic solvents, among others [90].

Laccase

Laccases catalyze reduction of O_2 to H_2O and can be applied to solve problems such as the presence of xenobiotic and aromatic compounds [92]. These enzymes are involved in the biodegradation of environmental pollutants, including the degradation of synthetic dyes as well as industrial applications in the areas of textiles and biofuels. Fungal laccases such as the one produced by *A. niger* completely discolor various synthetic dyes [93, 94].

Studies have suggested possibility of inserting laccases from *A. niger in situ for* biodegradation of low-density polyethylene, demonstrating a potential of 30% plastic weight loss, with the minimum of nutritional conditions for the development of the fungus and enzymatic production. In other words, even in adverse conditions, *A. niger* presented satisfactory results [95].

Table 1. Main conclusions, challenges and opportunities for *Aspergillus niger* enzymes production

Application	Aspergillus species	Main conclusions, challenges, and opportunities	Reference
Hydrolysis of raw starchy food stuff	*A. niger* RBP7	After fermentation the cell free supernatant was significantly effective for hydrolysis of taro, yam, malanga and sweet potato; Acidophilus amylase may be suitable for use to treat digestive dysfunction.	98
Baking processes	*A. niger* DP-Azb60	The food enzyme α-amylase from non-genetically modified *A. niger* strain did not give rise to safety concerns under the intended conditions of use.	99
Glucose syrups production, beverage alcohol and baking processes.	*A. niger* NZYM-MC	The food enzyme a-amylase from a genetically modified *A. niger* strain did not give rise to safety concerns.	100
Starch processing, beverage alcohol processes and baking processes	*A. niger* NZYM-SB	The food enzyme a-amylase from a genetically modified *A. niger* strain presented no safety concern for the environment.	101
Bioethanol	*A. niger*	The algal starch is a potential substrate for bioethanol production through simultaneous saccharification and fermentation; Future works must include the optimization of the saccharification.	102
Baking and starch industries	*A. niger* CBS513.88	A novel acid-tolerant α-amylase from *A. niger* was discovered and characterized, which could be used under acidic conditions.	19
Directly mining	*A. niger*	The study provides a direct way to mine active enzymes from man-made environment of Chinese Nong-flavor liquor starter.	103
Unspecified	*A. niger*	The highest activity unit for α-amylase and β-glucosidase were in 6 days of fermentation and using corn cob as substrate.	104
Unspecified	*A. niger*	*A. niger* is a good, cheap and easily available source to production of amylases from musk melon shells.	105
Starch industries	*A. niger*	Triticale grains without the addition of any nutritive components could be used as a sole substrate for enzymes production from fungus under SSF; The higher efficiency was in a wheat flour hydrolysis; The obtained *A. niger* amylase cocktail was more efficient than the commercially amylase cocktail.	106

The number of *A. niger* laccases is low; nevertheless, a study characterized the structure of these biocatalysts, revealing relevant biochemical differences, including the amino acid positions and the active site of the enzyme [96].

Among microbial enzymes, fungal enzymes have preference over other microbial sources because of their GRAS status (generally recognized as safe) [14]. Filamentous fungi are potential sources of enzyme production, considered even more stable than bacterial enzymes [16, 17].

The production of enzymes through fungi can be made from residues or byproducts from the agricultural industry such as sugarcane bagasse, banana and potato peels, wheat, corn, rice, soybean meal, among others [18]. Although these materials have polluting potential, they have viable uses as substrates for the growth of different microorganisms [97].

Table 1 shows opportunities and perspectives reported in the literature regarding the production of enzymes by *A. niger.*

ASPERGILLUS NIGER AND BIOPRODUCTS: INDUSTRIAL APPLICATIONS

In addition to enzymatic production, *A. niger* can produce other molecules and bioactive products of industrial interest. These important compounds can be called bioproducts. Some examples are organic acids, biosurfactants, biopeptides, biopolymers, among others [12, 107]. Bioproducts come from biomass, and their characteristics depend on the process of obtaining and separating biomass/bioproduct that ends up involving its cost in the market [108].

Compared to non-biological products, bioproducts offer important advantages, including biodegradability, efficiency in processing reactions under adverse pH and temperature conditions. Consequently, they have a good ability to act in reactions of interest to the food, textile, pharmaceutical, energy, among others [109].

Organic acids are widely used by the food industry, acting as acidulants, flavorings and preservatives. Using *A. niger* for fermentation with glucose or sucrose as substrates, two bioproducts of food interest are produced, citric and gluconic acids. Citric acid is widely used for pH control in jellies, fruit juice and wine; in addition, foods are treated with this acid for better retention of vitamin C [110]. Gluconic acid is a product of the conversion of glucose, that can be obtained by the enzymatic process; therefore, the process is more profitable, especially when the enzymes are produced microbially. This acid influences taste, aroma and food stability. Other fungi can be used to obtain these bioproducts such as *A. goneii* and *Penicillium citrinum*; nevetheless, *A. niger* is the most common and efficient [111].

Among biopolymers, microbial production of chitosan and chitin has applications in the areas of food and pharmaceuticals, acting as antioxidants, emulsifiers, thickeners, and stabilizers and in the production of dressings and anticoagulants [112]. Chitin and chitosan are structural components found in the cell wall of fungi. For this reason, they are excellent producers of these two components, and *A. niger* is considered a potential producer [113].

In the textile industry, biopolymers can also be used for the treatment of wastewater with a high load of dyes. Biopolymers produced using *A. niger* present an economical, effective and ecological treatment alternative, removing the dye by adsorption processes. This can be confirmed in studies that used *A. niger* to remove dye G yellow-thiazole, obtaining approximately 99.97% of removal of this pollutant at pH 6 [27]. The interesting thing is that the removal of other dyes such as azo red was also studied, and *A. niger* showed removal efficiency of about 58.6% [97].

From industrial processes to produce organic acids and enzymes, biomass residues are generated, and when *A. niger* is used, this residue can be used to obtain chitosan. More than 80,000 tons of *A. niger* biomass residues per year are generated only from citric acid production [113], considered a promising source.

The challenges for obtaining fungal chitosan focus on using alternative cultivation media for biomass production and the process of deacetylation of chitosan through temperature, time and concentration of milder basic compounds. According to Ghormade, Pathan and Deshpande [113], this would help minimize the environmental impacts of the process and increase the yield of this high added value bioproduct.

Biosurfactants have been gaining strength in the area of bioproducts, mainly because of the demand for biological products to replace chemicals, as they are less toxic and more biodegradable. These biocompounds have wide industrial applications, in cosmetics, pharmaceuticals, food and petroleum industries. They can also be applied as bioremediators of contaminated soil and water. In the food industry, their use was reported in emulsification processes and antimicrobial activity [114]. In the pharmaceutical industry, they have been reported in the production of antimicrobials such as echinocandin, a lipopeptide with antimicrobial activity responsible for combating diseases such as aspergillosis [115]. Regarding environmental applications, the use of this bioproduct applied to the treatment of contaminants is being advanced, where it can bioremediate compounds in soil and effluents, assisting in biodegradation mainly viaits media emulsification properties, increasing the bioavailability of the contaminants for microbial degradation [116]. Studies have shown that *A. niger* produces metabolites with characteristics of degrading hydrocarbons, even a consortium of bioemulsifying proteins [107, 117].

The production of biosurfactants and other biocompounds as lipolytic enzymes is also reported in the literature. Yaseen et al. [118] evaluated the potential of biosurfactants produced by *A. niger* to remediate lead-contaminated soil at 150 mg.kg^{-1}. The biosurfactant produced presented an emulsification index of 80% after 7 days of cultivation. In toxic metal bioremediation, after 15 days of experiment, the precipitation of the contaminant increased 30% when compared to the soil without treatment.

Proteins from various sources can undergo enzymatic hydrolysis during food processing or bioprocess, generating considerable impact on functional properties, improved nutritional value and high added value of the desired product. During protein hydrolysis, peptides of various lengths and free amino acids are generated as a result of cleavage of peptide bonds. Low-molecular-weight peptides can be absorbed directly into the intestine where they have minimal allergenic effects. In addition, these peptides have various bioactivities, including antioxidant, antihypertensive, anticarcinogenic, antimicrobial, immunomodulatory, antithrombotic, and diabetic control properties [119]. They behave like hormones or even as medicines, protecting or reducing the adverse effects of non-communicable chronic diseases [120]. The kind of proteases, microorganisms or methods adopted influence the size, sequences and types of amino acids, as well as the biological effects of the peptides.

Antifungal peptides are also secreted by filamentous fungi such as *A. niger* [121]. There are several studies on the protein activity of *A. niger*; however, there remains much to be explored regarding the biopeptides produced by the fungus, as well as their applicability and bioavailability after gastrointestinal digestion. The study of the simultaneous production of proteases and biopeptides is extremely important because the medium used as substrate can benefit from the hydrolytic capacity of fungal enzymes.

A. niger is considered an alternative and potential source for all bioproducts cited in this text, with the ability to produce a series of bioproducts with high added value that are beneficial for human, animal and environmental health; in short, it can be considered one of the primary microorganisms involved in the biotechnological revolution.

CHALLENGES AND PERSPECTIVES

Many bioproducts of technological interest mentioned in this review can also be produced by purely chemical reactions; nevertheless, such methods end up not being sufficiently economically competitive in comparison with the production routes from microbial fermentations [122]. For example, the commercial synthesis of citric acid by *A. niger* is superior to that of other microorganisms in terms of production yield, simplicity in handling and the ability to ferment various cheap raw materials [122, 123].

Several bioproducts are obtained from submerged fermentation, in which equipment and culture media are required at high cost and that also consume substantial amounts of energy. To meet challenges like this and the contamination risks that are associated with submerged fermentation, solid state fermentation has been shown to be a good strategy to allow better production of a wide range of enzymes and metabolites of biotechnological interest [124].

Multinational companies already use *A. niger* to produce citric acid and enzymes. Since the discovery of the microorganism more than 100 years ago, the scientific community has realized that *A. niger* is diverse and geographically dispersed. Several researchers seek strategies to insert this microorganism in relevant sectors such as bioremediation, geomicrobiology, pathogenicity, toxin production, food safety, agricultural microbiology, among others [12].

Studies show that the future of industrial biotechnology for *A. niger* over the next 20 years reserve a promising market in synthetic biology (with mycotoxin-free isolates), network analysis such as genomics, metabolomic gene expression, applications in co-cultivation, editing using the CRISPR-Cas9 system and greater interest in secondary metabolism, enzyme production, and citric acid, among others [12]. Among enzymes alone, the current market is 4.7 billion euros and this is projected to double by 2026 [1].

Strain selection and application of different mutagens, including ultraviolet radiation, X-rays and gamma rays, as well as chemicals such as ethyl methane sulfonate and diethyl sulfonate can be used to induce *A. niger* mutation for the creation of strains with potential industrial application for commercial production [122]. Genomic sequences of industrial strains of *A. niger* have increasingly become accessible to be better studied through comparative and functional analysis of the genome [125]. The ideal fungal genome should contain all genes sufficient for high productivity and robustness, eliminating all undesirable metabolic pathways or secondary clusters of metabolite genes that represent a major challenge in the industrial use of the microorganism [1].

Recent studies have reported that improvement in enzyme production occurred through changes in the morphology of *A. niger* using titanium microparticles. Two recombinant strains of *A. niger* that produce fructofuranosidase and glucoamylase were tested with titanite microparticles (TiSiO, 8 μm) in the growth medium in shake flasks to improve enzyme production. The study found that the improvement increased 3.7 times for fructofuranosidase and 9.5 times for glucoamylase. Such results can be used to create bio-pellets with high enzyme production in agitated tank reactors [126].

Advances such as those cited for the changes in *A. niger* biology could be useful in creating personalized organisms either to support specific production processes or to answer specific research questions [1]. The focus on systems biology and genome editing are two areas that have been growing in research in the last decade in the *A. niger* community [12]. Nevertheless, some of the remaining challenges consist in the analysis of *A. niger* genome sequences that need to be comprehensively resolved [122]. It was found that *A. niger* strains optimized by random mutagenesis maintained the capacity for the production of mycotoxins [127]. For example, many of the encoded genes have secreted peptides that hinder industrial protein production. Some (an estimated 10% of *A. niger* isolates used industrially) are capable of biosynthesizing mycotoxins such as fumosin and ochratoxin A (two toxic and carcinogenic agents) [1, 127]. It is very important that such mycotoxins not be present in the proteins used in human medicine. In some cases, the use of biomass resulting from the fermentation process for animal food is questionable for these reasons [127].

References

[1] Meyer V, Andersen MR, Brakhage AA, Braus GH, Caddick MX, Cairns TC, de Vries RP, Haarmann T, Hansen K, Hertz-Fowler C, Krappmann S, Mortensen UH, Peñalva MA, Ram AFJ, Head RM: Current Challenges of Research on Filamentous Fungi in Relation to Human Welfare and a Sustainable Bio-Economy: A White Paper. *Fungal Biology and Biotechnology* 2016, *3*(1), 6; https://doi.org/10.1186/s40694-016-0024-8.

[2] Schuster E, Dunn-Coleman N, Frisvad J, Dijck P van: On the Safety of *Aspergillus niger* - a Review. *Applied Microbiology and Biotechnology* 2002, *59*(4–5), 426–435; https://doi.org/10.1007/s00253-002-1032-6.

[3] Pandey A, Webb C, Fernandes M, Larroche C: *Enzyme Technology*, 1st ed.; Pandey, A., Webb, C., Fernandes, M., Larroche, C., Eds.; Springer: Verlag, New York, 2006.

[4] Pasin TM, Benassi VM, Heinen PR, Damasio AR de L, Cereia M, Jorge JA, Polizeli M de LT de M: Purification and Functional Properties of a Novel Glucoamylase Activated by Manganese and Lead Produced by Aspergillus Japonicus. *International Journal of Biological Macromolecules* 2017, *102*, 779–788; https://doi.org/10.1016/j.ijbiomac.2017.04.016.

[5] Martin MF, Okpo EA, Andy IE: Microbial Amylases : A Review. 2019, *22*(December 2018), 174–179.

[6] Colla LM, Rizzardi J, Pinto MH, Reinehr CO, Bertolin TE, Costa JAV: Simultaneous Production of Lipases and Biosurfactants by Submerged and Solid-State Bioprocesses. *Bioresource Technology* 2010, *101*(21), 8308–8314; https://doi.org/10.1016/j.biortech.2010.05.086.

[7] Gangadharan D, Sivaramakrishnan S, Nampoothiri KM, Sukumaran RK, Pandey A: Response Surface Methodology for the Optimization of Alpha Amylase Production by Bacillus Amyloliquefaciens. *Bioresource Technology* 2008, *99*(11), 4597–4602; https://doi.org/10.1016/j.biortech.2007.07.028.

[8] El-Bakry M, Abraham J, Cerda A, Barrena R, Ponsá S, Gea T, Sánchez A: From Wastes to High Value Added Products: Novel Aspects of SSF in the Production of Enzymes. *Critical Reviews in Environmental Science and Technology* 2015, *45*(18), 1999–2042; https://doi.org/10.1080/10643389.2015.1010423.

[9] Hölker U, Lenz J: Solid-State Fermentation — Are There Any Biotechnological Advantages? *Current Opinion in Microbiology* 2005, *8*(3), 301–306; https://doi.org/10.1016/j.mib.2005.04.006.

[10] Heerd D, Yegin S, Tari C, Fernandez-Lahore M: Pectinase Enzyme-Complex Production by Aspergillus Spp. in Solid-State Fermentation: A Comparative Study. *Food and Bioproducts Processing* 2012, *90*(2), 102–110; https://doi.org/10.1016/j.fbp.2011.08.003.

[11] Amin F,Bhatti HN,Bilal M: Recent Advances in the Production Strategies of Microbial Pectinases—A Review. *International Journal of Biological Macromolecules* 2019, *122*, 1017–1026; https://doi.org/10.1016/j.ijbiomac.2018.09.048.

[12] Cairns TC, Nai C, Meyer V: How a Fungus Shapes Biotechnology: 100 Years of *Aspergillus niger* Research. *Fungal Biology and Biotechnology* 2018, *5*(1), 13; https://doi.org/10.1186/s40694-018-0054-5.

[13] Vogel A, May O: *Industrial Enzyme Applications*; Eds.; John Wiley & Sons, Ltd: Chichester, UK, 2019; https://doi.org/10.1002/9783527813780.

[14] Gupta R, Gigras P, Mohapatra H, Goswami VK, Chauhan B: Microbial α-Amylases: A Biotechnological Perspective. *Process Biochemistry* 2003, *38*(11), 1599–1616; https://doi.org/10.1016/S0032-9592(03)00053-0.

[15] Avwioroko OJ, Anigboro AA, Unachukwu NN, Tonukari NJ: Isolation, Identification and in Silico Analysis of Alpha-Amylase Gene of *Aspergillus niger* Strain CSA35 Obtained from Cassava Undergoing Spoilage. *Biochemistry and Biophysics Reports* 2018, *14*, 35–42; https://doi.org/10.1016/j.bbrep.2018.03.006.

[16] Arotupin DJ, Fabunmi TB, Gabriel-Aj RAO: Enzyme Activity of Microorganisms Associated with Fermented Husk and Testa of Cola Acuminata. *Research Journal of Microbiology* 2015, *10*(10), 466–475; https://doi.org/10.3923/jm.2015.466.475.

[17] Almeida PZ de, Pereira MG, Carvalho CC de, Heinen PR, Ziotti LS, Messias JM, Jorge JA, Polizeli M de LT de M: Bioprospection and Characterization of the Amylolytic Activity by Filamentous Fungi from Brazilian Atlantic Forest. *Biota Neotropica* 2017, *17*(3); https://doi.org/10.1590/1676-0611-bn-2017-0337.

[18] Adejuwon AO, Tsygankova VA, Obayemi OS: α-Amylase Production Using Aspergilus Vadensis Isolated From Pulverized Cocoa Seeds. *Life Science Journal* 2019, *16*(8), 64–70; https://doi.org/10.7537/marslsj160819.08.

[19] Wang J, Zhang Y, Wang X, Shang J, Li Y, Zhang H, Lu F, Liu F: Biochemical Characterization and Molecular Mechanism of Acid Denaturation of a Novel α-Amylase from *Aspergillus niger*. *Biochemical Engineering Journal* 2018, *137*, 222–231; https://doi.org/10.1016/j.bej.2018.06.004.

[20] Gopinath SCB, Anbu P, Arshad MKM, Lakshmipriya T, Voon CH, Hashim U, Chinni S V.: Biotechnological Processes in Microbial Amylase Production. *BioMed Research International* 2017, *2017*, 1–9; https://doi.org/10.1155/2017/1272193.

[21] Steffen KT, Hofrichter M, Hatakka A: Mineralisation of 14 C-Labelled Synthetic Lignin and Ligninolytic Enzyme Activities of Litter-Decomposing Basidiomycetous Fungi. *Applied Microbiology and Biotechnology* 2000, *54*(6), 819–825; https://doi.org/10.1007/s002530000473.

[22] Rogalski J, Szczodrak J, Janusz G: Manganese Peroxidase Production in Submerged Cultures by Free and Immobilized Mycelia of Nematoloma Frowardii. *Bioresource Technology* 2006, *97*(3), 469–476; https://doi.org/10.1016/j.biortech.2005.03.002.

[23] Dhakar K, Kooliyottil R, Joshi A, Pandey A: Simultaneous Production of Ligninolytic Enzymes by a Temperature and PH Tolerant Strain of *Aspergillus niger* under Different Cultural Conditions. *Indian Journal of Biotechnology* 2015, *14*(1), 81–86.

[24] Jebor MA: *Production and Purification of Peroxidase* From. 2017, No. 2, 684–696.

[25] dos Santos TC, dos Santos Reis N, Silva TP, Pereira Machado F de P, Ferereira Bonomo RC, Franco M: Prickly Palm Cactus Husk as a Raw Material for Production of Ligninolytic Enzymes by *Aspergillus niger*. *Food Science and Biotechnology* 2016, *25*(1), 205–211; https://doi.org/10.1007/s10068-016-0031-9.

[26] Asses N, Ayed L, Hkiri N, Hamdi M: Congo Red Decolorization and Detoxification by *Aspergillus niger*: Removal Mechanisms and Dye Degradation Pathway. *BioMed Research International* 2018, *2018*, 1–9; https://doi.org/10.1155/2018/3049686.

[27] Bankole PO, Adekunle AA, Govindwar SP: Demethylation and Desulfonation of Textile Industry Dye, Thiazole Yellow G by *Aspergillus niger* LAG. *Biotechnology Reports* 2019, *23*, e00327; https://doi.org/10.1016/j.btre.2019.e00327.

[28] Li J, Wang J, Li J, Liu D, Li H, Gao W, Li J, Liu S: *Aspergillus niger* Enhance Bioactive Compounds Biosynthesis As Well As Expression of Functional Genes in Adventitious Roots of Glycyrrhiza Uralensis Fisch. *Applied Biochemistry and Biotechnology* 2016, *178*(3), 576–593; https://doi.org/10.1007/s12010-015-1895-5.

[29] Asemoloye MD, Ahmad R, Jonathan SG: Transcriptomic Responses of Catalase, Peroxidase and Laccase Encoding Genes and Enzymatic Activities of Oil Spill Inhabiting Rhizospheric Fungal Strains. *Environmental Pollution* 2018, *235*, 55–64; https://doi.org/10.1016/j.envpol.2017.12.042.

[30] El Enshasy HA, Elsayed EA, Suhaimi N, Malek RA, Esawy M: Bioprocess Optimization for Pectinase Production Using *Aspergillus niger* in a Submerged Cultivation System. *BMC Biotechnology* 2018, *18*(1), 71; https://doi.org/10.1186/s12896-018-0481-7.

[31] Sandri I,Silveira M: Production and Application of Pectinases from *Aspergillus niger* Obtained in Solid State Cultivation. *Beverages* 2018, *4*(3), 48; https://doi.org/10.3390/beverages4030048.

[32] Xia J, Li P: Pectic Enzymes. In *Encyclopedia of Food Chemistry*; Elsevier, 2019; pp 270–276; https://doi.org/10.1016/B978-0-08-100596-5.21645-4.

[33] Wang J, Chio C, Chen X, Su E, Cao F, Jin Y, Qin W: Efficient Saccharification of Agave Biomass Using *Aspergillus niger* Produced Low-Cost Enzyme Cocktail with Hyperactive Pectinase Activity. *Bioresource Technology* 2019, *272*, 26–33; https://doi.org/10.1016/j.biortech.2018.09.069.

[34] Patidar MK, Nighojkar S, Kumar A, Nighojkar A: Pectinolytic Enzymes-Solid State Fermentation, Assay Methods and Applications in Fruit Juice Industries: A Review. *3 Biotech* 2018, *8*(4), 199; https://doi.org/10.1007/s13205-018-1220-4.

[35] Gupta R, Ramnani P: Microbial Keratinases and Their Prospective Applications: An Overview. *Applied Microbiology and Biotechnology* 2006, *70*(1), 21–33; https://doi.org/10.1007/s00253-005-0239-8.

[36] Sharma R, Devi S: Versatility and Commercial Status of Microbial Keratinases: A Review. *Reviews in Environmental Science and Bio/Technology* 2018, *17*(1), 19–45; https://doi.org/10.1007/s11157-017-9454-x.

[37] de Oliveira CC, de Souza AKS, de Castro RJS: Bioconversion of Chicken Feather Meal by *Aspergillus niger*: Simultaneous Enzymes Production Using a Cost-Effective Feedstock Under Solid State Fermentation. *Indian Journal of Microbiology* 2019, *59*(2), 209–216; https://doi.org/10.1007/s12088-019-00792-3.

[38] Ai O, Cic O: *Food Science and Nutrition Technology Keratin Degradation by Aspergillus niger and Penicillium Purpurogenum Isolated from Nigerian Soil.* 2019, 1–5; https://doi.org/10.23880/fsnt-16000173.

[39] Mazotto AM, Couri S, Damaso MCT, Vermelho AB: Degradation of Feather Waste by *Aspergillus niger* Keratinases: Comparison of Submerged and Solid-State Fermentation. *International Biodeterioration & Biodegradation* 2013, *85*, 189–195; https://doi.org/10.1016/j.ibiod.2013.07.003.

[40] Lopes FC, Silva LAD e, Tichota DM, Daroit DJ, Velho RV, Pereira JQ, Corrêa APF, Brandelli A: Production of Proteolytic Enzymes by a Keratin-Degrading *Aspergillus niger*. *Enzyme Research* 2011, *2011*, 1–9; https://doi.org/10.4061/2011/487093.

[41] Umedum C, Ugorji I, Okoye E, Egemonye O, Obiora O: Keratinase Activity and Biodegradation Properties of Hyphomycetous Fungi from Fowl Feather. *International Journal of Agriculture and Biosciences* 2013, *2*(2), 306–309.

[42] Chen X, Zhou B, Xu M, Jia G, Huang Z, Zhao H, Liu G: Prokaryotic Expression and Characterization of a Keratinolytic Protease from *Aspergillus niger*. *Biologia* 2015, *70*(2); https://doi.org/10.1515/ biolog-2015-0031.

[43] Payne CM, Knott BC, Mayes HB, Hansson H, Himmel ME,Sandgren M, Ståhlberg J, Beckham GT: Fungal Cellulases. *Chemical Reviews* 2015, *115*(3), 1308–1448; https://doi.org/10.1021/cr500351c.

[44] Siqueira JGW, Rodrigues C, Vandenberghe LP de S, Woiciechowski AL, Soccol CR: Current Advances in On-Site Cellulase Production and Application on Lignocellulosic Biomass Conversion to Biofuels: A Review. *Biomass and Bioenergy* 2020, *132*, 105419; https://doi.org/10.1016/j.biombioe.2019.105419.

[45] Camargo AF, Scapini T, Stefanski FS, Kubeneck S, Preczeski KP, Venturin B, Bonatto C, Bordin ER, Müller C, Fongaro G, Treichel H: Reducing the Water Footprint of Bioethanol. In *Ethanol as a Green Alternative Fuel: Insight and Perspectives*; 2019; pp 199–220.

[46] Liang L, Xue D: Kinetics of Cellulose Hydrolysis by Halostable Cellulase from a Marine *Aspergillus niger* at Different Salinities. *Process Biochemistry* 2017, *63*, 163–168; https://doi.org/10.1016/j.procbio.2017.09.004.

[47] Cai LN, Xu SN, Lu T, Lin DQ, Yao SJ: Salt-Tolerant Mechanism of Marine *Aspergillus niger* Cellulase Cocktail and Improvement of Its Activity. *Chinese Journal of Chemical Engineering* 2019; https://doi.org/10.1016/j.cjche.2019.11.012.

[48] Xue D, Lin D, Gong C, Peng C, Yao S: Expression of a Bifunctional Cellulase with Exoglucanase and Endoglucanase Activities to Enhance the Hydrolysis Ability of Cellulase from a Marine *Aspergillus niger*. *Process Biochemistry* 2017, *52*, 115–122; https://doi.org/10.1016/j.procbio.2016.09.030.

[49] Hu Y, Du C, Pensupa N, Lin CSK: Optimisation of Fungal Cellulase Production from Textile Waste Using Experimental Design. *Process*

Safety and Environmental Protection 2018, *118*, 133–142; https://doi.org/10.1016/j.psep.2018.06.009.

[50] Zhao C, Deng L, Fang H: Mixed Culture of Recombinant Trichoderma Reesei and *Aspergillus niger* for Cellulase Production to Increase the Cellulose Degrading Capability. *Biomass and Bioenergy* 2018, *112*, 93–98; https://doi.org/10.1016/j.biombioe.2018.03.001.

[51] Idris ASO, Pandey A, Rao SS, Sukumaran RK: Cellulase Production through Solid-State Tray Fermentation, and Its Use for Bioethanol from Sorghum Stover. *Bioresource Technology* 2017, *242*, 265–271; https://doi.org/10.1016/j.biortech.2017.03.092.

[52] Mulyaningtyas A, Sediawan WB: Effect of Combined Pretreatment of Lignocellulose and the Kinetics of Its Subsequent Bioconversion by *Aspergillus niger*. *Biocatalysis and Agricultural Biotechnology* 2019, *21*, 101292; https://doi.org/10.1016/j.bcab.2019.101292.

[53] Wang J, Chen X, Chio C, Yang C, Su E, Jin Y, Cao F, Qin W: Delignification Overmatches Hemicellulose Removal for Improving Hydrolysis of Wheat Straw Using the Enzyme Cocktail from *Aspergillus niger*. *Bioresource Technology* 2019, *274*, 459–467; https://doi.org/10.1016/j.biortech.2018.12.029.

[54] Ahmad R, Khare SK: Immobilization of *Aspergillus niger* Cellulase on Multiwall Carbon Nanotubes for Cellulose Hydrolysis. *Bioresource Technology* 2018, *252*, 72–75; https://doi.org/10.1016/j.biortech.2017.12.082.

[55] Raza S, Yong X, Deng J: Immobilizing Cellulase on Multi-Layered Magnetic Hollow Particles: Preparation, Bio-Catalysis and Adsorption Performances. *Microporous and Mesoporous Materials* 2019, *285*, 112–119; https://doi.org/10.1016/j.micromeso.2019.05.011.

[56] Sarsaiya S, Awasthi SK, Awasthi MK, Awasthi AK, Mishra S, Chen J: The Dynamic of Cellulase Activity of Fungi Inhabiting Organic Municipal Solid Waste. *Bioresource Technology* 2018, *251*, 411–415; https://doi.org/10.1016/j.biortech.2017.12.011.

[57] Martinelle M, Holmquist M, Hult K: On the Interfacial Activation of Candida Antarctica Lipase A and B as Compared with Humicola Lanuginosa Lipase. *Biochimica et Biophysica Acta (BBA) - Lipids and Lipid Metabolism* 1995, *1258*(3), 272–276; https://doi.org/10.1016/0005-2760(95)00131-U.

[58] Sharma R, Chisti Y, Banerjee UC: Production, Purification, Characterization, and Applications of Lipases. *Biotechnology Advances* 2001, *19*(8), 627–662; https://doi.org/10.1016/S0734-9750(01)00086-6.

[59] Tan JS, Abbasiliasi S, Ariff AB, Ng HS, Bakar MHA, Chow YH: Extractive Purification of Recombinant Thermostable Lipase from Fermentation Broth of Escherichia Coli Using an Aqueous Polyethylene Glycol Impregnated Resin System. *3 Biotech* 2018, *8*(6), 288; https://doi.org/10.1007/s13205-018-1295-y.

[60] Kamini NR, Mala JGS, Puvanakrishnan R: Lipase Production from *Aspergillus niger* by Solid-State Fermentation Using Gingelly Oil Cake. *Process Biochemistry* 1998, *33*(5), 505–511; https://doi.org/10.1016/S0032-9592(98)00005-3.

[61] Cordova J, Nemmaoui M, Ismaïli-Alaoui M, Morin A, Roussos S, Raimbault M, Benjilali B: Lipase Production by Solid State Fermentation of Olive Cake and Sugar Cane Bagasse. *Journal of Molecular Catalysis B: Enzymatic* 1998, *5*(1–4), 75–78; https://doi.org/10.1016/S1381-1177(98)00067-8.

[62] Mala JGS, Edwinoliver NG, Kamini NR, Puvanakrishnan R: Mixed Substrate Solid State Fermentation for Production and Extraction of Lipase from *Aspergillus niger* MTCC 2594. *The Journal of General and Applied Microbiology* 2007, *53*(4), 247–253; https://doi.org/10.2323/jgam.53.247.

[63] Treichel H, Sbardelotto M, Venturin B, Agnol AD, Mulinari J, Golunski SM, Baldoni DB, Bevilacqua CB, Seminoti Jacques RJ, Vargas GDLP, Mossi AJ: Lipase Production from a Newly Isolated *Aspergillus niger* by Solid State Fermentation Using Canola Cake as Substrate. *Current Biotechnology* 2017, *6*(4); https://doi.org/10.2174/2211550105666151124193225.

[64] Reinehr CO, Treichel H, Tres MV, Steffens J, Brião VB, Colla LM: Successive Membrane Separation Processes Simplify Concentration of Lipases Produced by *Aspergillus niger* by Solid-State Fermentation. *Bioprocess and Biosystems Engineering* 2017, *40*(6), 843–855; https://doi.org/10.1007/s00449-017-1749-3.

[65] Preczeski KP, Kamanski AB, Scapini T, Camargo AF, Modkoski TA, Rossetto V, Venturin B, Mulinari J, Golunski SM, Mossi AJ, Treichel H: Efficient and Low-Cost Alternative of Lipase Concentration Aiming at the Application in the Treatment of Waste Cooking Oils. *Bioprocess and Biosystems Engineering* 2018, *41*(6), 851–857; https://doi.org/10.1007/s00449-018-1919-y.

[66] Chen C, Tian H, Xing S, Li C, Zeng X, He L: Influence of Different Parameters on Reverse Micelle Extraction Combined with Acetone Precipitation to Purify Sn-1,3 Extracellular Lipase from *Aspergillus niger* GZUF36. *Journal of Food Science and Technology* 2019, *56*(6), 2899–2908; https://doi.org/10.1007/s13197-019-03743-4.

[67] Nadar SS, Rathod VK: Encapsulation of Lipase within Metal-Organic Framework (MOF) with Enhanced Activity Intensified under Ultrasound. *Enzyme and Microbial Technology* 2018, *108*, 11–20; https://doi.org/10.1016/j.enzmictec.2017.08.008.

[68] Feng K, Huang Z, Peng B, Dai W, Li Y, Zhu X, Chen Y, Tong X, Lan Y, Cao Y: Immobilization of *Aspergillus niger* Lipase onto a Novel Macroporous Acrylic Resin: Stable and Recyclable Biocatalysis for Deacidification of High-Acid Soy Sauce Residue Oil. *Bioresource Technology* 2020, *298*, 122553; https://doi.org/10.1016/j.biortech.2019.122553.

[69] Hu Y, Dai L, Liu D, Du W: Rationally Designing Hydrophobic UiO-66 Support for the Enhanced Enzymatic Performance of Immobilized Lipase. *Green Chemistry* 2018, *20*(19), 4500–4506; https://doi.org/10.1039/C8GC01284A.

[70] Prabaningtyas RK, Putri DN, Utami TS, Hermansyah H: Production of Immobilized Extracellular Lipase from *Aspergillus niger* by Solid State Fermentation Method Using Palm Kernel Cake, Soybean

Meal, and Coir Pith as the Substrate. *Energy Procedia* 2018, *153*, 242–247; https://doi.org/10.1016/j.egypro.2018.10.010.

[71] Utami TS, Hariyani I, Alamsyah G, Hermansyah H: Production of Dry Extract Extracellular Lipase from *Aspergillus niger* by Solid State Fermentation Method to Catalyze Biodiesel Synthesis. *Energy Procedia* 2017, *136*, 41–46; https://doi.org/10.1016/j.egypro. 2017. 10.275.

[72] Guldhe A, Singh P, Kumari S, Rawat I, Permaul K, Bux F: Biodiesel Synthesis from Microalgae Using Immobilized *Aspergillus niger* Whole Cell Lipase Biocatalyst. *Renewable Energy* 2016, *85*, 1002–1010; https://doi.org/10.1016/j.renene.2015. 07.059.

[73] Mandari V, Nema A, Devarai SK: Sequential Optimization and Large Scale Production of Lipase Using Tri-Substrate Mixture from *Aspergillus niger* MTCC 872 by Solid State Fermentation. *Process Biochemistry* 2019; https://doi.org/10.1016/j.procbio.2019.10.026.

[74] Ajijolakewu AK, Leh CP, Wan Abdullah WN, Lee C keong: Optimization of Production Conditions for Xylanase Production by Newly Isolated Strain *Aspergillus niger* through Solid State Fermentation of Oil Palm Empty Fruit Bunches. *Biocatalysis and Agricultural Biotechnology* 2017, *11*, 239–247; https://doi.org/10. 1016/j.bcab.2017.07.009.

[75] Li Q, Wu T, Duan Y, Pei J, Zhao L: Improving the Thermostability and PH Stability of *Aspergillus niger* Xylanase by Site-Directed Mutagenesis. *Applied Biochemistry and Microbiology* 2019, *55*(2), 136–144; https://doi.org/10.1134/S0003683819020108.

[76] Podestá MV, Morilla EA, Allasia MB, Woitovich Valetti N, Tubio G, Boggione MJ: An Eco-Friendly Method of Purification for Xylanase from *Aspergillus niger* by Polyelectrolyte Precipitation. *Journal of Polymers and the Environment* 2019, *27*(12), 2895–2905; https://doi.org/10.1007/s10924-019-01571-3.

[77] Amorim CC, Farinas CS, Miranda EA: Liquefied Wheat Bran as Carbon Source and Inducer in High-Solids Submerged Cultivation of *Aspergillus niger* for Xylanase Production. *Biocatalysis and*

Agricultural Biotechnology 2019, *21*, 101346; https://doi.org/10.1016/j.bcab.2019.101346.

[78] Sridevi A, Sandhya A, Ramanjaneyulu G, Narasimha G, Devi PS: Biocatalytic Activity of *Aspergillus niger* Xylanase in Paper Pulp Biobleaching. *3 Biotech* 2016, *6*(2), 165; https://doi.org/10.1007/s13205-016-0480-0.

[79] Elegbede JA, Lateef A, Azeez MA, Asafa TB, Yekeen TA, Oladipo IC, Aina DA, Beukes LS, Gueguim-Kana EB: Biofabrication of Gold Nanoparticles Using Xylanases Through Valorization of Corncob by *Aspergillus niger* and Trichoderma Longibrachiatum: Antimicrobial, Antioxidant, Anticoagulant and Thrombolytic Activities. *Waste and Biomass Valorization* 2018; https://doi.org/10.1007/s12649-018-0540-2.

[80] C. Aragon C, I. Ruiz-Matute A, Corzo N, Monti R, M. Guisán J,Mateo C: Production of Xylo-Oligosaccharides (XOS) by Controlled Hydrolysis of Xylan Using Immobilized Xylanase from *Aspergillus niger* with Improved Properties. *Integrative Food, Nutrition and Metabolism* 2018, *5*(4); https://doi.org/10.15761/IFNM.1000225.

[81] Souza PM de, Bittencourt ML de A, Caprara CC, Freitas M de, Almeida RPC de, Silveira D, Fonseca YM, Ferreira Filho EX, Pessoa Junior A, Magalhães PO: A Biotechnology Perspective of Fungal Proteases. *Brazilian Journal of Microbiology* 2015, *46*(2), 337–346; https://doi.org/10.1590/S1517-838246220140359.

[82] Snyman C, Theron L, Divol B: Understanding the Regulation of Extracellular Protease Gene Expression in Fungi: A Key Step towards Their Biotechnological Applications. *Applied Microbiology and Biotechnology* 2019, *103*(14), 5517–5532; https://doi.org/10.1007/s00253-019-09902-z.

[83] McCotter SW, Horianopoulos LC, Kronstad JW: Regulation of the Fungal Secretome. *Current Genetics* 2016, *62*(3), 533–545; https://doi.org/10.1007/s00294-016-0578-2.

[84] Pel HJ, de Winde JH, Archer DB, Dyer PS, Hofmann G, Schaap PJ, Turner G, de Vries RP, Albang R, Albermann K, Andersen MR,

Bendtsen JD, Stam H, et al.: Genome Sequencing and Analysis of the Versatile Cell Factory *Aspergillus niger* CBS 513.88. *Nature Biotechnology* 2007, *25*(2), 221–231; https://doi.org/10.1038/nbt1282.

[85] Vishwanatha KS, Rao AGA, Singh SA: Acid Protease Production by Solid-State Fermentation Using *Aspergillus oryzae* MTCC 5341: Optimization of Process Parameters. *Journal of Industrial Microbiology & Biotechnology* 2010, *37*(2), 129–138; https://doi.org/10.1007/s10295-009-0654-4.

[86] Sandhya C, Sumantha A, Szakacs G, Pandey A: Comparative Evaluation of Neutral Protease Production by *Aspergillus oryzae* in Submerged and Solid-State Fermentation. *Process Biochemistry* 2005, *40*(8), 2689–2694; https://doi.org/10.1016/j.procbio.2004.12.001.

[87] Hajji M, Rebai A, Gharsallah N, Nasri M: Optimization of Alkaline Protease Production by *Aspergillus clavatus* ES1 in Mirabilis Jalapa Tuber Powder Using Statistical Experimental Design. *Applied Microbiology and Biotechnology* 2008, *79*(6), 915; https://doi.org/10.1007/s00253-008-1508-0.

[88] Purushothaman K, Bhat SK, Singh SA, Marathe GK, Appu Rao ARG: Aspartic Protease from *Aspergillus niger*: Molecular Characterization and Interaction with Pepstatin A. *International Journal of Biological Macromolecules* 2019, *139*, 199–212; https://doi.org/10.1016/j.ijbiomac.2019.07.133.

[89] Sattar H, Bibi Z, Kamran A, Aman A, Ul Qader SA: Degradation of Complex Casein Polymer: Production and Optimization of a Novel Serine Metalloprotease from *Aspergillus niger* KIBGE-IB36. *Biocatalysis and Agricultural Biotechnology* 2019, *21*, 101256; https://doi.org/10.1016/j.bcab.2019.101256.

[90] de Castro RJS, Soares MH, Albernaz JRM, Sato HH: Biochemical Characterization of Solvent, Salt, Surfactant and Oxidizing Agent Tolerant Proteases from *Aspergillus niger* Produced in Different Agroindustrial Wastes. *Biocatalysis and Agricultural Biotechnology* 2016, *5*, 94–98; https://doi.org/10.1016/j.bcab.2015.12.011.

[91] Scherf KA, Wieser H, Koehler P: Novel Approaches for Enzymatic Gluten Degradation to Create High-Quality Gluten-Free Products. *Food Research International* 2018, *110*, 62–72; https://doi.org/10.1016/j.foodres.2016.11.021.

[92] Antecka A, Blatkiewicz M, Bizukojć M, Ledakowicz S: Morphology Engineering of Basidiomycetes for Improved Laccase Biosynthesis. *Biotechnology Letters* 2016, *38*(4), 667–672; https://doi.org/10.1007/s10529-015-2019-6.

[93] Iqbal HMN, Kyazze G, Tron T, Keshavarz T: Laccase from *Aspergillus niger* : A Novel Tool to Graft Multifunctional Materials of Interests and Their Characterization. *Saudi Journal of Biological Sciences* 2018, *25*(3), 545–550; https://doi.org/10.1016/j.sjbs.2016.01.027.

[94] Copete-Pertuz LS, Alandete-Novoa F, Plácido J, Correa-Londoño GA, Mora-Martínez AL: Enhancement of Ligninolytic Enzymes Production and Decolourising Activity in Leptosphaerulina Sp. by Co–Cultivation with Trichoderma Viride and *Aspergillus terreus*. *Science of The Total Environment* 2019, *646*, 1536–1545; https://doi.org/10.1016/j.scitotenv.2018.07.387.

[95] Sáenz M,Borodulina T,Diaz L,Banchon C: Minimal Conditions to Degrade Low Density Polyethylene by *Aspergillus terreus* and *niger*. *Journal of Ecological Engineering* 2019, *20*(6), 44–51; https://doi.org/10.12911/22998993/108699.

[96] Ferraroni M, Westphal AH, Borsari M, Tamayo-Ramos JA, Briganti F, Graaff LH de, Berkel WJH van: Structure and Function of *Aspergillus niger* Laccase McoG. *Biocatalysis* 2017, *3*(1), 1–21; https://doi.org/10.1515/boca-2017-0001.

[97] Mahmoud MS, Mostafa MK, Mohamed SA, Sobhy NA, Nasr M: Bioremediation of Red Azo Dye from Aqueous Solutions by A *Spergillus niger* Strain Isolated from Textile Wastewater. *Journal of Environmental Chemical Engineering* 2017, *5*(1), 547–554; https://doi.org/10.1016/j.jece.2016.12.030.

[98] Mukherjee R, Paul T, Soren JP, Halder SK, Mondal KC, Pati BR, Das Mohapatra PK: Acidophilic α-Amylase Production from

Aspergillus niger RBP7 Using Potato Peel as Substrate: A Waste to Value Added Approach. *Waste and Biomass Valorization* 2019, *10*(4), 851–863; https://doi.org/10.1007/s12649-017-0114-8.

[99] Silano V, Barat Baviera JM, Bolognesi C, Brüschweiler BJ, Cocconcelli PS, Crebelli R, Gott DM, Grob K, Lampi E, Mortensen A,Rivière G, Steffensen I, Tlustos C, Van Loveren H, Vernis L, Zorn H, Jany K, Glandorf B, Penninks A, Želježic D, Andryszkiewicz M, Arcella D, Liu Y, Rossi A, Engel K, Chesson A: Safety Evaluation of the Food Enzyme Alpha-amylase from Non-genetically Modified *Aspergillus niger* Strain (Strain DP-Azb60). *EFSA Journal* 2019, *17*(5); https://doi.org/10.2903/j. efsa.2019. 5680.

[100] Silano V, Barat Baviera JM, Bolognesi C,Brüschweiler BJ, Cocconcelli PS, Crebelli R, Gott DM, Grob K, Lampi E, Mortensen A, Riviere G, Steffensen I, Tlustos C, van Loveren H, Vernis L, Zorn H, Kärenlampi S, Marcon F, Penninks A, Smith A, Aguilera-Gómez M, Andryszkiewicz M, Arcella D, Kovalkovičová N, Liu Y, Rossi A, Engel K, Chesson A: Safety Evaluation of the Food Enzyme A-amylase from a Genetically Modified *Aspergillus niger* (Strain NZYM-MC). *EFSA Journal* 2018, *16*(10); https://doi.org/ 10.2903/j.efsa.2018.5451.

[101] Silano V, Bolognesi C, Castle L, Chipman K, Cravedi J, Fowler P, Franz R, Grob K, Gürtler R, Husøy T, Kärenlampi S, Mennes W, Milana MR, Pfaff K, Riviere G, Srinivasan J, Tavares Poças M de F, Tlustos C, Wölfle D, Zorn H, Chesson A, Glandorf B, Herman L, Jany K, Marcon F, Penninks A, Smith A, Van Loveren H, Želježić D, Aguilera-Gómez M, Andryszkiewicz M, Arcella D, Kovalkovičová N, Liu Y, Engel K: Safety Evaluation of the Food Enzyme A-amylase from a Genetically Modified *Aspergillus niger* (Strain NZYM-SB). *EFSA Journal* 2018, *16*(7); https://doi.org/10. 2903/j.efsa.2018.5320.

[102] Singh S, Chakravarty I, Pandey KD, Kundu S: Development of a Process Model for Simultaneous Saccharification and Fermentation

(SSF) of Algal Starch to Third-Generation Bioethanol. *Biofuels* 2018, 1–9; https://doi.org/10.1080/17597269.2018.1426162.

[103] Yi Z, Fang Y, He K, Liu D, Luo H, Zhao D, He H, Jin Y, Zhao H: Directly Mining a Fungal Thermostable α-Amylase from Chinese Nong-Flavor Liquor Starter. *Microbial Cell Factories* 2018, *17*(1), 30; https://doi.org/10.1186/s12934-018-0878-y.

[104] Aliyah A, Alamsyah G, Ramadhani R, Hermansyah H: Production of α-Amylase and β-Glucosidase from *Aspergillus niger* by Solid State Fermentation Method on Biomass Waste Substrates from Rice Husk, Bagasse and Corn Cob. *Energy Procedia* 2017, *136*, 418–423; https://doi.org/10.1016/j.egypro.2017.10.269.

[105] Rao MRK, Kumar S, Kumar N: Production of amylase from cucumis melo using *Aspergillus niger* by liquid fermentation. *Indo American Journal of Pharmaceutical Sciences* 2017, *4*(12), 4584–4589.

[106] Dojnov B, Grujic M, Vujcic Z: Highly Efficient Production of *Aspergillus niger* Amylase Cocktail by Solid-State Fermentation Using Triticale Grains as a Well-Balanced Substrate. *Journal of the Serbian Chemical Society* 2015, *80*(11), 1375–1390; https://doi.org/10.2298/JSC150317041D.

[107] Cerda A, Artola A, Barrena R, Font X, Gea T, Sánchez A: Innovative Production of Bioproducts From Organic Waste Through Solid-State Fermentation. *Frontiers in Sustainable Food Systems* 2019, *3*; https://doi.org/10.3389/fsufs.2019.00063.

[108] Budzianowski WM: High-Value Low-Volume Bioproducts Coupled to Bioenergies with Potential to Enhance Business Development of Sustainable Biorefineries. *Renewable and Sustainable Energy Reviews* 2017, *70*, 793–804; https://doi.org/10.1016/j.rser.2016.11.260.

[109] Jemli S, Ayadi-Zouari D, Hlima H Ben, Bejar S: Biocatalysts: Application and Engineering for Industrial Purposes. *Critical Reviews in Biotechnology* 2016, *36*(2), 246–258; https://doi.org/10.3109/07388551.2014.950550.

[110] Steiger MG,Rassinger A,Mattanovich D,Sauer M: Engineering of the Citrate Exporter Protein Enables High Citric Acid Production in *Aspergillus niger*. *Metabolic Engineering* 2019, *52*, 224–231; https://doi.org/10.1016/j.ymben.2018.12.004.

[111] Copetti MV: Fungi as Industrial Producers of Food Ingredients. *Current Opinion in Food Science* 2019, *25*, 52–56; https://doi.org/10.1016/j.cofs.2019.02.006.

[112] Abdel-Gawad KM, Hifney AF, Fawzy MA, Gomaa M: Technology Optimization of Chitosan Production from *Aspergillus niger* Biomass and Its Functional Activities. *Food Hydrocolloids* 2017, *63*, 593–601; https://doi.org/10.1016/j.foodhyd.2016.10.001.

[113] Ghormade V, Pathan EK, Deshpande MV: Can Fungi Compete with Marine Sources for Chitosan Production? *International Journal of Biological Macromolecules* 2017, *104*, 1415–1421; https://doi.org/10.1016/j.ijbiomac.2017.01.112.

[114] Gharibzahedi SMT, Hernández-Ortega C, Welti-Chanes J, Putnik P, Barba FJ, Mallikarjunan K, Escobedo-Avellaneda Z, Roohinejad S: High Pressure Processing of Food-Grade Emulsion Systems: Antimicrobial Activity, and Effect on the Physicochemical Properties. *Food Hydrocolloids* 2019, *87*, 307–320; https://doi.org/10.1016/j.foodhyd.2018.08.012.

[115] Mnif I, Ghribi D: Review Lipopeptides Biosurfactants: Mean Classes and New Insights for Industrial, Biomedical, and Environmental Applications. *Biopolymers* 2015, *104*(3), 129–147; https://doi.org/10.1002/bip.22630.

[116] Lin L, Cunshan Z, Vittayapadung S, Xiangqian S, Mingdong D: Opportunities and Challenges for Biodiesel Fuel. *Applied Energy* 2011, *88*(4), 1020–1031; https://doi.org/10.1016/j.apenergy.2010.09.029.

[117] Sánchez-Vázquez V,Shirai K,González I,Gutiérrez-Rojas M: Polycyclic Aromatic Hydrocarbon-Emulsifier Protein Produced by *Aspergillus brasiliensis* (*niger*) in an Airlift Bioreactor Following an Electrochemical Pretreatment. *Bioresource Technology* 2018, *256*, 408–413; https://doi.org/10.1016/j.biortech.2018.02.043.

[118] Yaseen RY, Abd El-Aziz SM, Eissa DT, Abou-Shady AM: Application of Biosurfactant Producing Microorganisms to Remediate Heavy Metal Pollution in El-Gabal El-Asfar Area. *Alexandria Science Exchange Journal: An International Quarterly Journal of Science Agricultural Environments* 2018, *39* (January-March), 17–34; https://doi.org/10.21608/asejaiqjsae.2018.5504.

[119] Sila A, Bougatef A: Antioxidant Peptides from Marine By-Products: Isolation, Identification and Application in Food Systems. A Review. *Journal of Functional Foods* 2016, *21*, 10–26; https://doi.org/10.1016/j.jff.2015.11.007.

[120] Zanutto-Elgui MR, Vieira JCS, Prado DZ do, Buzalaf MAR, Padilha P de M, Elgui de Oliveira D, Fleuri LF: Production of Milk Peptides with Antimicrobial and Antioxidant Properties through Fungal Proteases. *Food Chemistry* 2019, *278*, 823–831; https://doi.org/10.1016/j.foodchem.2018.11.119.

[121] Gun Lee D, Yub Shin S, Maeng CY, Zhu Jin Z, Lyong Kim K, Hahm KS: Isolation and Characterization of a Novel Antifungal Peptide from *Aspergillus niger*. *Biochemical and Biophysical Research Communications* 1999, *263*(3), 646–651; https://doi.org/10.1006/bbrc.1999.1428.

[122] Show PL, Oladele KO, Siew QY, Aziz Zakry FA, Lan JCW, Ling TC: Overview of Citric Acid Production from *Aspergillus niger*. *Frontiers in Life Science* 2015, *8*(3), 271–283; https://doi.org/10.1080/21553769.2015.1033653.

[123] Themelis DG, Tzanavaras PD: Reagent-Injection Spectrophotometric Determination of Citric Acid in Beverages and Pharmaceutical Formulations Based on Its Inhibitory Effect on the Iron(III) Catalytic Oxidation of 2,4-Diaminophenol by Hydrogen Peroxide. *Analytica Chimica Acta* 2001, *428*(1), 23–30; https://doi.org/10.1016/S0003-2670(00)01221-6.

[124] Nema A, Patnala SH, Mandari V, Kota S, Devarai SK: Production and Optimization of Lipase Using *Aspergillus niger* MTCC 872 by Solid-State Fermentation. *Bulletin of the National Research Centre* 2019, *43*(1), 82; https://doi.org/10.1186/s42269-019-0125-7.

[125] Li X, Xie X, Liu J, Wu D, Cai G, Lu J: Characterization of a Putative Glycoside Hydrolase Family 43 Arabinofuranosidase from *Aspergillus niger* and Its Potential Use in Beer Production. *Food Chemistry* 2020, *305*, 125382; https://doi.org/10.1016/j.foodchem.2019.125382.

[126] Driouch H, Hänsch R, Wucherpfennig T, Krull R, Wittmann C: Improved Enzyme Production by Bio-Pellets of *Aspergillus niger*: Targeted Morphology Engineering Using Titanate Microparticles. *Biotechnology and Bioengineering* 2012, *109*(2), 462–471; https://doi.org/10.1002/bit.23313.

[127] Frisvad JC, Larsen TO, Thrane U, Meijer M, Varga J, Samson RA, Nielsen KF: Fumonisin and Ochratoxin Production in Industrial *Aspergillus niger* Strains. *PLoS ONE* 2011, *6*(8), e23496; https://doi.org/10.1371/journal.pone.0023496.

In: *Aspergillus niger*
Editor: Eva Baughan

ISBN: 978-1-53618-080-0

Chapter 3

ASPERGILLUS NIGER AND CITRIC ACID

T. Ramesh[1,*], H. Ann Suji[1], M. Thangaraj[1], R. Kumaran[2], D. Annadurai[1] and M. Shenbagam[3]

[1]Centre of Advanced Study in Marine Biology,
Faculty of Marine Sciences, Annamalai University, Parangipettai,
Tamilnadu, India

[2]Department of Zoology,
Dharmapuram Gnanambigai Govt. Arts College (W), Mayiladuthurai,
Tamilnadu, India

[3]Department of Biochemistry and Biotechnology,
Annamalai University, Annamalainagar, India

ABSTRACT

Aspergillus niger is one of the most significant fungi that is comprehensively utilized in different natural fields. It has been utilized for a long time in the creation of extracellular enzymes and citric acid. Over the last two decades, *A. niger* has been created as a significant change host to overexpress nourishment enzymes. *A. niger* strains produce a progression of secondary metabolites by fermentation that has

* Corresponding Author's Email: rameshcasmb@gmail.com.

demonstrated to be an important biomaterial. From a natural perspective, it very well may be utilized as a bioadsorbent to detoxify and decolorize the wastewaters. In industrial and pharmaceutical applications, it can similarly be utilized as a cheap and attractive hotspot for business creation of some costly and helpful biopolymers, such as chitin, chitosan, and nourishment enhancements like chitin-glucan and glucosamine. This section is, for the most part, centred around the development of citric acid by *A. niger*.

Keywords: heterotrophic fungi, *Aspergillus niger*, fermentation, citric acid, agricultural wastes

INTRODUCTION

Biosynthetic creation of chemical compounds by microorganisms causes an extraordinary positive change in the overall piece of the pie. Among them, organic acids establish a critical job as a result of their useful structures, which are the beginning material for chemical industries. Organic acids are concoction mixes that are generally circulated in nature as typical constituents of plants or creature tissues. Microbial generation of organic acids is a promising methodology (Sauer et al., 2008). It is amazing that such huge numbers of fungi have a high proficiency of organic corrosive generation at high focuses. Filamentous fungi are economically utilized for creation of various organic acids (Gupta et al., 2018). *Aspergillus niger* can discharge a mixed drink of hydrolytic proteins to quickly debase complex polymers found in modest substrates (for example, feedstock) into monosaccharides (Tong et al., 2019).

They are created through the digestion of starches (Dubey, 2001) that are either the terminal products of Embden-Meyerhof-Parnas (EMP) pathway (Glycolysis). Examples include lactic acid and propionic acid or the products of incomplete oxidation of sugars (citric acid, itaconic acid, and gluconic acid). A third type of item is likewise received from the dehydrogenation of alcohol within the sight of oxygen. An example of this would be acetic acid (Riviere, 1997). They offer an extraordinary potential for industrial use because they are fabricated on an enormous scale. They are advertised generally as unadulterated chemicals or their particular salts.

Among the natural acids, citric acid is the most significant business item that is delivered in a huge scale. Chemically, citric acid is named as 2-hydroxy-1, 2, 3-propanetricarboxylic acid, and this is found in practically all plant and animal tissues. It generally exists in nature and presents as an organic product acid in lemon, orange, pineapple, plum, peas, and peach. It is also present in animal bones, muscles, and blood. It was first detached from lemon and lime squeezes by Scheele in 1784 (King and Cheetham, 1987). It was not until 1920 that a business beneficial strategy appeared which involved making citric acid from a hot lemon squeeze by the sedimentation of calcium carbonate.

WORLD MARKET DEMAND IN CITRIC ACID PRODUCTION

Based on the report of the following research article, the global citric acid utilization was estimated at 67% in 2016, (https://www.marketreports world.com/enquiry/request-sample/13221907) and this will only increase by 2024.

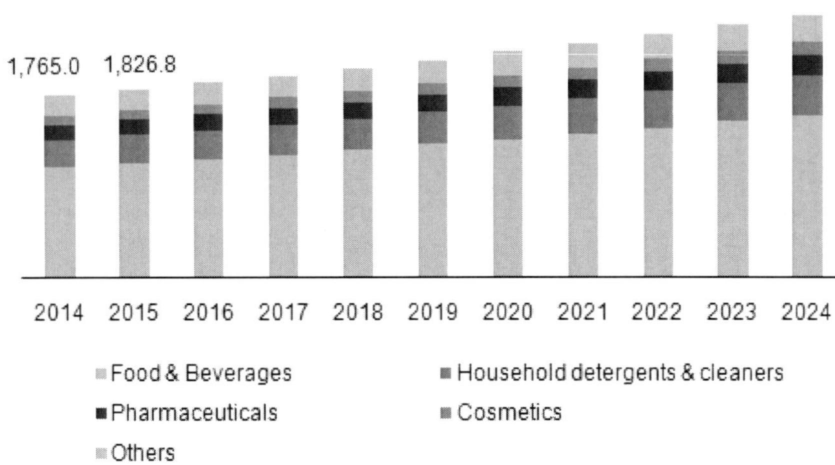

Courtesy: Market Research Report, Hexa Research

Global citric acid market volume, 2014–2024 (Kilo Tons).

To satisfy the need, it is important to follow alternative methods of microbial production of citric acid. This is simple these days because of the innovative improvements in microbial alterations through Genomics, Transcriptomics, metabolic engineering, quantitative proteomics, and various other methods. Owing to novel discoveries in industrial production, systems such as progressed downstream processing techniques and process engineering ensure the chance of profitable production in chemical industries. Paul et al. (2019) recommended that DNA sequencing, Recombinant DNA technology, and Microarray techniques alongside gene assembling methods through Restriction enzyme-based methods, Restriction enzyme-based scarless methods, Sequence homology-based methods, and Homology recombination-based methods were reasonable for expanding the production of wanted chemicals.

DEVELOPMENTS IN CITRIC ACID PRODUCTION

In 1893, Wehmer detailed in his investigation that the culture of *Penicillium glaucum* can create citric acid by fermentation on sugar. At that point, he found that two new fungal strains that consisted of a similar capacity. Soon enough, he disengaged the two new fungal strains with the capacity to collect the citric acid, and they were assigned as *Citromyces (Penicillium)*. Issues such as contamination and the lengthy duration of fermentation hampered these procedures to be successful at a commercial level. In 1917, Currie built up a form fermentation that made the industrialization of citric acid production conceivable. In 1919, a Belgian baker prevailed in the shallow dish fermentation process. In 1923, the most useful fermentation for production of citric acid started, utilizing microorganisms developing in surface culture. Fermentation utilizing deep-vat fermenters started in the 1930s in certain parts of England, Soviet Union, and Germany (Berovic and Legisa, 2007). Rohr et al. (1983) revealed that citric acid can be created by the fermentation process. This was done by utilizing different types of microorganisms, in particular *Aspergillus niger*, which is a filamentous fungus most commonly used in business production, because of its cultural characteristics such as development at low pH and high sugar concentration.

BIOCHEMISTRY OF CITRIC ACID (CA)

CA is an essential metabolic item and is framed in the tricarboxylic acid (TCA) cycle. Glucose is the principle carbon source utilized for citric acid creation. In numerous living beings, 80% of the glucose is separated by responses of EMP pathway and 20% by response of Pentose–Phosphate pathway for citric acid biosynthesis.

Citric acid has been created in three phases:

1. The breakdown of hexoses to pyruvate and acetyl-CoA by glycolysis
2. Oxaloacetic acid (Oxaloacetate) formation
3. Reduction of acetyl-CoA and oxaloacetic acid into citric acid

These conversions lead to a sufficient quantity of oxaloacetic acid to aid continuous production. The revival of oxaloacetic acid is done by four mechanisms:

1. The direct carboxylation of pyruvate is catalyzed by malic enzyme. It provides malate, which is readily oxidized into oxaloacetate through malic dehydrogenase.
2. The carboxylation of pyruvate is catalyzed by pyruvate carboxylase.

$$\text{Pyruvate} + CO_2 + H_2O + \text{ATP} \underset{}{\overset{Mg^{++}}{\rightleftarrows}} \text{Oxaloacetate} + \text{ADP} + \text{Pi}$$

3. The carboxylation of phosphoenol pyruvate (PEP) is catalyzed by PEP carboxykinase.

$$PEP + ADP + CO_2 \rightleftharpoons Oxaloacetate + ATP$$

4. Through the glyoxylate cycle involving the key enzymes isocitratelyase and malate synthase

Isocitrate → Succinate + Glyoxylate
↓
Acetyl-CoA + Glyoxylate ⇌ Malate + Co-enzyme A
↓
Citric acid ← Oxaloacetate

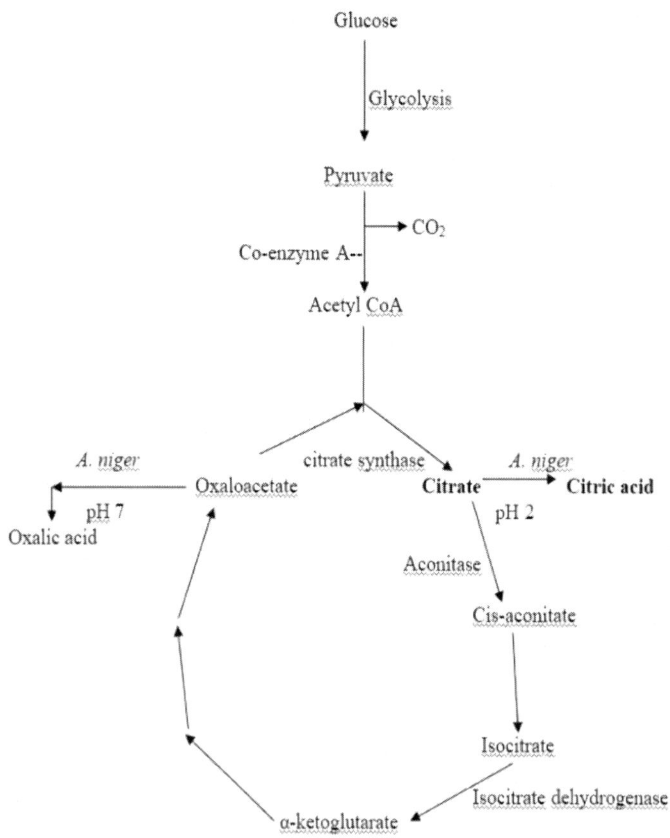

Figure 1. Citric acid production in the TCA Cycle.

MICROBES INVOLVED

There are numerous microorganisms, including bacteria, yeast, and fungi, that have been employed for citric acid production (Soccol et al., 2006). Examples include *Arthrobacter paraffinens*, *Bacillus licheniformis* and *Corynebacterium* ssp., *Aspergillus niger*, *A. aculeatus*, *A. carbonarius*, *A. awamori*, *A. foetidus*, *A. fonsecaeus*, *A. phoenicis*, *Penicillium janthinellum*, *Candida tropicalis*, *C. oleophila*, *C. guilliermondii*, *C. citroformans*, *Hansenula anamola*,and *Yarrowia lipolytica*. Even though various organisms are associated with citric acid creation, *A. niger* is the best and reasonable parasites in mechanical generation because of the simplicity, right from its capacity to ferment an assortment of modest raw materials to the significant returns it provides with relatively simple care. Modern strains that produce commercial citric acid are not unreservedly accessible, and only a couple can be acquired from universal culture collections.

Aspergillus niger

Aspergillus niger is one of the most widely recognized species of mold. It is also called black mold and is seen in grapes, onions, and other different crisp nourishments. Despite the fact that mold isn't as liable to present high risk to human health as different strands of *Aspergillus*, it may still cause aspergillosis (a lung disease) and otomycosis (fungal ear infection).

CLASSIFICATION

The taxonomic grouping of *Aspergillus niger* recognizes the fungus in the species *A. niger*, genus *Aspergillus*, family Trichocomaceae, order Eurotiales, phylum Ascomycota, and inside the kingdom of Fungi. There are 15 other similar fungi that are normally mistaken for *A. niger*. While all these represent a similar danger to humans, *A. niger* is well on the way to begin a fungal infection.

IDENTIFICATION

Aspergillus niger is at first white; however, as it develops, it turns black on the top and pale yellow at the bottom. The fungus is flexible, ready to flourish in different conditions, and has a delicate surface. All fungi inside the *Aspergillus* species develop rapidly, so early detection is imperative to forestall health issues.

Toxic Substances Control Act (TSCA, 1997) gives the last risk assessment of *A. niger* in industrial fermentations. *Aspergillus niger* is a worldwide in appropriation and has been confined from various territories. Humans are persistently presented to *A. niger* through spores and vegetative forms that is available in food and is noticeable almost everywhere. By far, most of the strains of *A. niger*, particularly those used in industrial fermentations, have a past filled with safe use. Keeping in mind that there are sporadic reports, despite what might be expected, most cases have not been recorded as serious pathogens of humans, animals, or plants. Explicit strains may deliver certain mycotoxins or may trigger allergic responses among workers. Those restricted occasions of unfriendly impacts are, by all accounts, related characterizations of industrial strains, and the use of those with the potential for such an impact can be avoided.

An unconstrained ascent in intracellular cAMP levels was seen during the nascent periods of *A. niger* development under high citric acid yielding conditions. The measure of cAMP shaped was subject to introductory sucrose concentration. The intracellular citric acid had begun to gather earlier; all the more quickly, it may be that cAMP union is activated by intracellular acidification caused by dissociation of citric acid (Grapulin and Legisa, 1997).

CITRIC ACID PRODUCTION TECHNIQUES

Chemical synthesis of citric acid is conceivable, yet it is no less expensive than fungal fermentation. Be that as it may, a modest quantity of citric acid, under 1% of complete world creation, is still derived from citrus fruits in Mexico and South America, where citrus fruits are accessible financially. Lately, citric acid has been created by fungal fermentation. There are essentially three unique sorts of batch fermentation process utilized in industries. These are the Japanese Koji process, the liquid surface culture process, and the submerged fermentation process.

In 1952, America Miles Company embraced the deep-level fermentation technique to create citric acid in a large scale. In the 1970s, the meteoric rise of various research and development centers across the world has helped many countries advance. Therefore, the complete yield of citric acid generation expanded quickly. Toward the end of 1978, there were 35 countries delivering citric acid with a yield of 40 thousand tons, which jumped to 500 thousand tons in 1995. A few specialists have contemplated continuous fermentation on a laboratory scale. However, no business generation by this sort of process is known (Yigitoglu, 1992).

Fermentation is the best, most reasonable, and most utilized method in citric acid production. In addition, it is a basic and steady system that gives more yield than different strategies; the prerequisites required for the production are relatively few and comprise effortlessly-controlled parameters that favor circumstances in the fermentation process. The fermentation process is conveyed in two significant stages. The first is the upstream or the inoculum preparation process, and the second is the downstream or product recovery process. With the assistance of raw materials, production medium preparation and fungal spores inoculation are the principal phases of upstream processing fermentation; in the subsequent stage, they are under controlled and optimized parameters such as pH, temperature, incubation period, aeration, and so on. The fermentation process is completed in the fermentor. In the last stage, the citric acid is recouped from the production medium by filtration, centrifugation, and separation procedures. A few types of fermenters have been utilized for citric acid production in solid-state fermentation. Examples include Erlenmeyer-tapered jars, glass incubators, trays, rotating and horizontal drum bioreactors, packed-bed segment bioreactors, single-layer packed-bed bioreactors, multi-layer packed-bed bioreactors, and so on.

SUBMERGED FERMENTATION (SMF)

After the selection of an organism, the production process must be developed. Submerged fermentation has been broadly utilized for mechanical production. Submerged fermentation alludes to those fermentations wherein the microorganisms utilized develop in a submerged state inside the fermentation media. In SmF, free-flowing liquid substrates such as molasses and broths are utilized. The finished results of the fermentation are freed into the fermentation broth. Substrate utilization is fast in SmF; thereafter, substrate must be given continuously to the fermentation process. This system is appropriate for the extraction of the metabolites from the bacteria because of the high-moisture content requirement for their growth (Couto and Sanromán, 2006). SmF has a few points of interest where genetically modified organisms are grown well compared to SSF and media sterilization, purification, and recovery of the finished results. Further, the control of process parameters, for example, pH, temperature, moisture, oxygen transfer, and aeration, should be possible effectively. (Kunamneni et al., 2005).

Bioreactors for submerged fermentation are created all around, and the parameters are controlled online. In particular, mass transfer and heat removal are controlled effectively today. About, 80% of the world citrus extract generation is derived from submerged fermentation. The primary focal points are higher productivity and yields, lower work costs, lower contamination risk, and work consumption.

Submerged fermentation equipment is ordinarily intended for monoseptic cultures and is equipped for being sterilized by steam under pressure. The significant types of submerged fermenters are the mixed tank fermenter, bubble section, carrier fermenter, fluidized-bed fermenter, and stream-bed fermenter.

The liquid medium is utilized in submerged fermentation, which stays in contact with the microorganism. An appropriate inventory of aeration is required for submerged fermentation. There are four types of culture techniques accessible for the growth of these bioreactors: batch culture, fed-batch culture, perfusion batch culture, and continuous culture. In batch culture, the microbes are inoculated into the fixed volume of medium. On account of fed-batch culture, the concentrated parts of the supplement are added bit by bit to the batch culture. In perfusion batch culture, the expansion of the culture and withdrawal of an equivalent volume of utilized cell-free medium is performed. In continuous culture, fresh medium is included into the batch framework during the exponential phase of microbial growth with a corresponding withdrawal of the medium containing the item. Continuous development gives close-adjusted growth, with little change in nutrients, metabolites, cell numbers, or biomass.

EXTRACTION

The broth is filtered until it is clear. Calcium citrate is precipitated by the expansion of magnesium-free $Ca(OH)_2$. When compared to calcium, magnesium is more soluble. It leads to the formation of magnesium citrate in the production medium. Because of this, magnesium-free calcium hydroxide is used. Calcium citrate is filtered, and calcium is precipitated with the addition of sulfuric acid to the filter cake. The activated carbon is added to the diluted solution for purifying the citric acid and passing through iron exchange beds. The obtained solution is allowed to evaporate to yield crystals of citric acid. Further purification might be required to meet pharmaceutical stipulations.

RECOVERY

In submerged fermentation, the subsequent stage after the end of fermentation is the partition of microbial cells and suspended solids from the liquid medium. This is generally accomplished by filtration for fungi or by centrifugation for bacteria and yeast.

SOLID-STATE FERMENTATION (SSF) OR KOJI PROCESS

This system was first evolved in Japan, where copious raw materials such as fruit squanders and rice bran, are accessible. This is the least difficult technique, and it has been an elective strategy for utilizing agro-industrial deposits. Solid-state fermentation is appropriate for the microorganisms that require low moisture content. In SSF, nutrient-rich waste materials, such as bran, bagasse, and paper pulp, can be utilized as a substrate for the microorganisms, and they are devoured gradually and continually. Subsequently, there is no compelling reason to supply the substrate for a longer time (Kunamneni et al., 2005). A few similitudes are seen with the surface process since the fungus additionally develops on a material surface. The substrate is solid, and it is dampened to around 70% moisture, contingent upon the substrate-absorption capacity. SSF includes a solid substrate on which both microbial growth and product amalgamation happen without free water (Pandey, 2003). The underlying pH of the material is typically adjusted to 4.5–6.0, and the temperature of incubation is around 28–30°C, contingent upon the microorganism utilized. The solid culture process is completed within 96 hours under optimal conditions. The most well-known organism utilized in solid-state fermentation is *A. niger* and there has been a similar report with yeasts. The strains with huge prerequisites of nitrogen and phosphorus are not perfect microorganisms for solid culture because of the lower diffusion rate of nutrients and metabolites happening at lower water activity in solid-state process. The nearness of trace elements may not influence citric acid production so hurtfully as it does in submerged fermentation; for the same reason, substrate pretreatment isn't required. This is one of the significant favorable circumstances of the solid culture (Soccol et al., 2006).

SSF is favorable in that it guarantees higher productivity, low energy requirements, economic possibility, upgraded product recovery, and creates less wastewater in comparison to SmF, making it a more favored strategy (El Aty et al., 2014). The fundamental motivation behind SSF is to get the fungi or bacteria in close contact with the insoluble substrate so that they can get the greatest nutrition from the substrate for fermentation (Bhargav et al., 2008). SSF has been seen as more appropriate for fungi and yeast as their water activity requirements (aw 0.5–0.6) are lower than that of bacteria (aw 0.8–0.9) (Thomas et al., 2013).

RAW MATERIALS

Soccol et al. (2006) arranged the raw materials that were utilized as the substrates for citric acid production by submerged, semisolid production (Table 1and Table2) and solid-state production of citric acid by different *A. niger* strains. The losses of agricultural, cellulosic materials such as cotton squanders were utilized by Kiel et al. (1981) rather than, or in addition to, sugar for citric acid production. Hossain et al. (1984) demonstrated the idea that the sugar source strongly affects citric acid production, and they found the exercises of specific enzymes. The impact of different carbon sources and their concentration on the production of citrate by *A. niger* has been examined by Xu et al. (1989). They found that the sugars maltose, sucrose, glucose, mannose, and fructose were utilized as carbon sources and gave exceptional returns of citric acid.

Legisa and Grapulin (1995) depicted that *A. niger* had the option to germinate and grow in substrates containing fundamentally higher beginning sucrose concentrations, wherein glycerol acts as an intracellular osmoregulator. At the point when citric acid began to be discharged into the medium, the substrate was unexpectedly diluted. Shu and Johnson (1947) reported the nearness of Mn and malt extract in the sporulation medium decreased the citric acid production in submerged fermentation, but not in the surface culture fermentation.

Rodrigues et al. (2013) upgraded the citric acid production by forced aeration method, in their experiment, they applied aeration by diffusion method. This method upgraded the cultures of *A. niger* strains CCT 7716 (parental strain), CCT 7718, and CCT 7717 (mutant strain), which improve the citric acid production by utilizing agro-industrial waste as a solid substrate.

Table 1. Raw materials employed in submerged and semi-solid production of citric acid

Raw material	Strain	Yield%
Beet molasses	*A. niger* ATTC 9142 *Yarrowia lipolytica* A101	- 68.7[a]
Black strap molasses	*A. niger* GCM 7	-
Brewery wastes	*A. niger* ATTC 9142	78.5
Cane molasses	*A. niger* T 55 *A. niger* GCMC-7	65 100
Carob pod extract	*A. niger*	
Coconut oil	*Candida. lipolytica* N-5704	99.6[b]
Corn starch	*A. niger* IM-155	62
Date syrup	*A. niger* ATTC 9142	50
Glycerol	*C. lipolytica* N–5704	58.8[b]
Hydrolysate starch	*Y. lipolytica* DS-1 *Y. lipolytica* A-101 *A. niger* UE-1	– 75 49
n-paraffin	*C. lipolytica* N-5704	161[b]
Olive oil	*C. lipolytica* N-5704	119[b]
Palm oil	*C. lipolytica* N-5704	155[b]
Rapeseed oil	*Y. lipolytica* A-101 *A. niger*	57 115[b]
Soybean oil	*Y. lipolytica* A-101	63
Soybean oil	*C. lipolytica* N-5704	115[b]
Wood hemicelluloses	*A. niger* IMI-41874 *S. lipolytica* IFO 1658	45[a] 41
Xylanhydrolysate	*A. niger* YANG No. 2	–
Yam bean starch	*A. niger* YW-112	74[a]

[a]based on sugar consumed; [b]based on oils and fatty acids

(The table is adapted and modified from the Soccol et al. (2006))

Table 2. Raw materials employed in solid-state production of citric acid

Raw material	Strain	Yield %
Apple pomace	*A. niger* NRRL 2001 *A. niger* NRRL 2270 *A. niger* NRRL 599 *A. niger* NRRL 328 *A. niger* NRRL 567 *A. niger* BC1	– – – – – 80
Carob pod extract	*A. niger* ATCC 9142	60
Carrot waste	*A. niger* NRRL 2270	36
Cassava bagasse Flasks Semi-pilot scale	*A. niger* LPB-21 *A. niger* LPB-21 *A. niger* CFTRI 30	 67 –
Cellulose hydrolysate and sugar cane	*A. niger*	44
Coffee husk	*A. niger* CFTRI 30	–
Corncob	*A. niger* NRRL 2001 *A. niger* NRRL 2270	50
Deoiled rice bran	*A. niger* CFTRI 30	–
Grape pomace	*A. niger* NRRL 2001 *A. niger* NRRL 2270 *A. niger* NRRL 599 *A. niger* NRRL 328 *A. niger* NRRL 567	88 – – – –
Kiwifruit peel	*A. niger* NRRL 567	–
Kumara (starch containing)	*A. niger* YANG No.2	–
Molasses (sugarcane bagasse)	*A. niger* DS1 Clarified Non-clarified molasses	 64.5 62.5
Mussel processing Wastes (polyurethane foams)	*A. niger*	–
Okara (soy residue)	*A. niger*	53
Orange waste	*A. niger*	–
Pineapple waste	*A. niger* ATCC 1015 *A. niger* ACM 4942	– 74
Rice bran	*A. niger* CFTRI 30	–
Sucrose (sugar cane bagasse)	*A. niger* CFTRI 30	–
Sugarcane-pressmud and wheat bran (4:1	*A. niger* CFTRI 30	
Wheat bran	*A. niger* CFTRI 30	–

[a]based on sugar consumed; [b]based on dry matter

As in every single other procedure where citric acid is made, the fermenter is made of acid-resistant materials, such as stainless steel. The carbohydrate sources are molasses decationized by ion exchange, sucrose, or glucose. $MgSO_4$, $7H_2O$, and KH_2PO_4 at about 1% and 0.05–2% individually are included (submerged fermentation phosphate restriction is a bit much). The pH is never allowed to rise above 3.5. Copper is utilized at up to 500 ppm as an antagonist of the enzyme aconitase, which requires iron. 1–5% of methanol, isopranol, or ethanol, when added to fermentations containing unpurified materials, increment the yield; the yields are decreased in media with purified materials. As high aeration is deleterious to citric acid production, mechanical agitation is a bit much and air might be bubbled through. Anti-form is included. The fungus happens as a uniform dispersal of pellets in the medium. The fermentation goes on for five to fourteen days.

SURFACE FERMENTATION

In surface fermentation, well-grown microbes appeared on the surface of the fermentation media. The model for the citric acid surface fermentation process wherein the microorganism *A. niger* grows as a thick, floating mycelial mat over the surface of the media utilized. It was the primary strategy followed by the industries in citric acid production and is still in use in small and medium-scale industries because of its affordability and simplicity. Gupta and Sharma (1994) obtained the best return of citric acid production by using combined cultures of immobilized and surface stabilized *A. niger* KCU 520 in continuous citric acid production from sugarcane molasses. In their test, before feeding into the reactor, the fermentation medium was presaturated with sterile air. Moreover, with surface aeration and this upgrade, 91% of the yield was obtained at the production rate of 1.7 kg M^{-3} h^{-1} with the concentration of 110kg M^{-3}. They utilized this culture for 30 days for citric acid production with no loss of productivity.

Ramesh and Kalaiselvam (2011) elaborately contemplated the citric acid production by utilizing red seaweed as an alternate marine substrate by *A. niger*. In their trial, *A. niger* was isolated from the marine soil sample and distinguished by lactophenol cotton blue (LCB) mount strategy. The seaweed was gathered from the coastal area in Rameswaram and was brought to the laboratory; it was identified by the keys. It was shade-dried for several days and then powdered mechanically. Seaweed powder was tried for its antifungal activity against *A. niger*. Carbohydrate estimation was carried out by adhering to standard strategy. From the 7-day old fungal culture, a loop-full of fungal spores was transferred to 5 ml of the sterile Potato dextrose broth and kept in a rotary shaker for the formation of pellets. The production medium was prepared, and the initial pH was adjusted to 6.5. At that point, the medium was sterilized at 15lbs pressure and at 121°C for 15 minutes. After sterilization, the medium was allowed to cool at room temperature. At that point, the pellet type of fungal culture was transferred and blended in the sterilized production medium. The inoculated flask was placed into a water bath shaker at 30°C and 200 rpm for 10 days for appropriate aeration. The pH was checked in aseptic conditions every 2 days.

PRODUCTION OF CITRIC ACID USING SEAWEED POWDER

Three trials were completed to discover the potential outcomes of utilizing seaweed as a substitute substrate for CA production. The control production medium (Table 3) was kept up in the main examination; in the subsequent trial, sucrose was supplanted by the equivalent volume of crude seaweed powder; and in the third investigation, 10% of sucrose was added to the crude seaweed powder in the fermentation medium for watching the improved activity of seaweed. The fermentation conditions were kept up cautiously all through the procedure.

Table 3. Composition of the production medium

Composition	Quantity
Sucrose	140.00 g
Ammonium nitrate	2.23 g
Dipotassiumortho phosphate	1.00 g
Magnesium sulphate	0.23 g
Distilled water	1000 ml
pH	6.5
Trace amount of Cu and Fe ions	

Note: Sucrose replaced by crude seaweed powder in the second experiment and crude seaweed powder (126 g) and 10% sucrose (14 g) in the third experiment

(Courtesy: Ramesh and Kalaiselvam, 2011)

Preparation of Inoculum

From the 7-day old fungal culture, a loop-full of fungal spores was transferred to 5 ml of the sterile Potato dextrose broth.

Conditions for Fermentation

According to Haider (2014) and Kudzai et al. (2016), citric acid production is primarily influenced by cultural conditions, such as carbon sources and concentration, nitrogen sources and concentration, and acidity of the medium and aeration. In building up a fermentation process for acquiring a valuable product, the optimization of cultivation conditions and determination of suitable substrates and their concentrations have essential importance because of their effect on the economy and achievability of the process. Numerous examinations have been planned to acquire more significant returns by utilizing optimization processes with minimal effort (Almousa et al., 2018).

The factors principally influencing the citric fermentation are the sort and concentration of carbon source, nitrogen and phosphate limitation, pH, aeration, oligoelements concentration, and morphology of the producing microorganism. Certain nutrients must be in over abundance, (for example, sugars, protons, or oxygen), others at limiting levels, (for example, nitrogen and phosphate), and others underneath settled limit esteems (for example, follow metals such as manganese) (Max et al., 2010; Nitin et al., 2017; Guc and Erkmen, 2017).

The production medium was readied, and the initial pH was acclimated to 6.5. Then, the medium was sterilized at 15lbs pressure and 121°C for 15 minutes. After sterilization, the medium was permitted to cool at room temperature. Then, a loop-full of fungal mycelium was moved from the inoculum and blended in the sterilized production medium. The inoculated flask was set into a water bath shaker at 30°C and 200 rpm for 10 days for proper aeration. The pH was checked in aseptic conditions every two days.

RECOVERY PROCESS

The stages involved in citric acid recuperation were as follows:

1. After the consummation of fermentation process, the incubated broth was filtered through the Whatman filter paper No. 8 for the detachment of fungal biomass and the fermentation broth. This broth goes about as the wellspring of citric acid.
2. Lime $(Ca\ (OH))_2$ was added to the fermentation broth to permit precipitation of citric acid in the form of calcium citrate.
3. Again, the precipitate was treated with diluted sulphuric acid to precipitate insoluble calcium sulfate. It was then filtered.
4. The precipitated arrangement containing citric acid was purified by passing through a column of carbon granules. Prior to using the carbon granules, they were treated with heat.
5. The obtained arrangement was evaporated under vacuum condition for getting the purified citric acid in crystal form.

CITRIC ACID DETERMINATION

The crystallized citric acid (CA) powder was determined titrimetrically using 0.1 NaOH and phenolphthalein as the indicator. The percentage (%) of CA was calculated using the given formula:

$$\% \text{ CA} = \frac{\text{Normality x Volume of NaOH x Equivalent weight of CA}}{\text{Weight of sample (g) x 10}}$$

Normality of NaOH used for titration is 0.1 N.
Volume of NaOH is the amount of NaOH that gives the color change.
Equivalent weight of citric acid is 48.03.
Weight is sample, which is the amount of sample taken for titration.

CITRIC ACID PRODUCTION

Fermentation process was completed for CA creation. During the fermentation procedure, the temperature was maintained at 30°C until the end of the procedure. There was a continuous decrease of pH (Table 4) seen in all the examinations, and it indicated the creation of acid.

Table 4. The pH variations at regular intervals of experimental period when seaweed powder of *G. acerosa* is inoculated with *A. niger*

Incubation period (days)	pH in production medium	pH in seaweed in corporated production medium	pH in seaweed+ 10% of sucrose in the production medium
2	6.0	6.5	6.0
4	5.0	6.0	6.0
6	3.5	5.5	5.0
8	3.5	5.0	4.0
10	1.5	3.5	3.0

Note: Initial pH is 6.5 at 0 day of incubation.
(Courtesy: Ramesh and Kalaiselvam, 2011)

In the first experiment, the initial pH was reduced on the second day (6.0) onward, and on the tenth day, the pH was at 1.5 in production medium; whereas in the production medium with seaweed and without sucrose, the pH was reduced from 6.5 on the second day to 3.5 on the tenth day. In the third experiment where the seaweed was supplemented with 10% sucrose in the medium, the pH reduced from 6.0 (second day) to 3.0 (tenth day) (Figure 2).

CITRIC ACID DETERMINATION

Citric acid was recovered in crystallized form from the fermentation broth following standard procedure. It was titrated with NaOH using phenolphthalein as the indicator; the colourless indicator changed into violet red, and the percentage of citric acid was calculated (Figure 3).

After recovery process, a purified citric acid was obtained, and the amount of yield is given in Table 5.

The percentage of CA produced in the different production media is given in Figure 4.

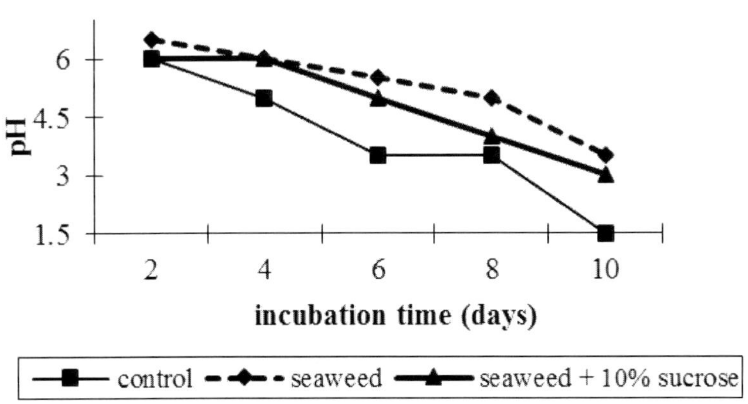

(Courtesy: Ramesh and Kalaiselvam, 2011)

Figure 2. pH reduction at regular intervals.

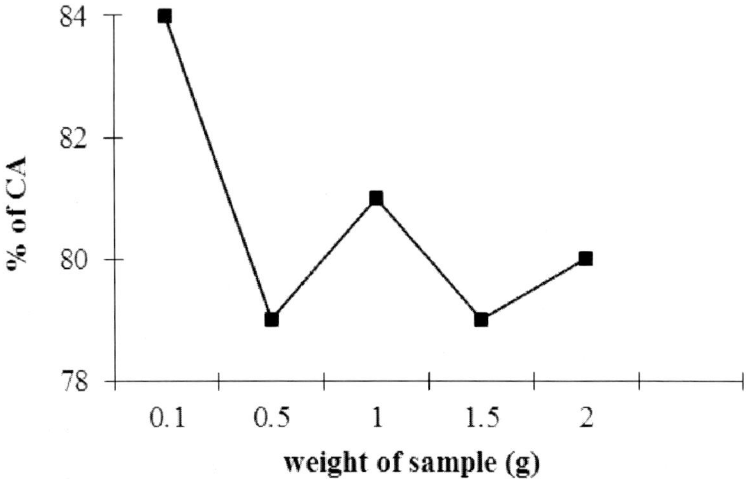

(Courtesy: Ramesh and Kalaiselvam, 2011)

Figure 3. Percentage (%) of citric acid.

Table 5. The yield and percentage of citric acid in different media

Type of the medium (without seaweed)	Yield (g L^{-1})
Production medium	80
Seaweed in the production medium	30
Seaweed + 10% of sucrose in the production medium	50

(Courtesy: Ramesh and Kalaiselvam, 2011)

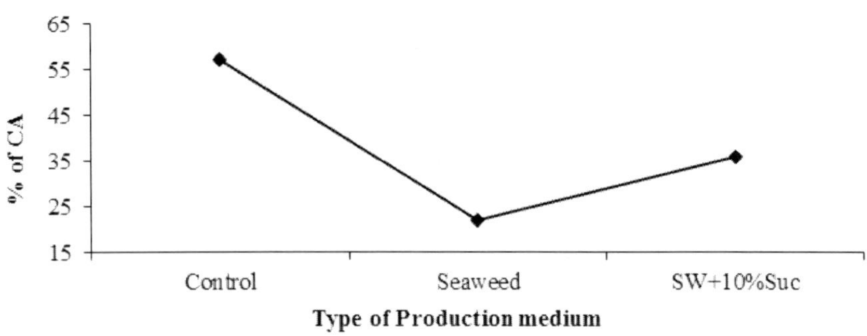

(Courtesy: Ramesh and Kalaiselvam, 2011)

Figure 4. Percentage (%) of CA production in different types of media.

The economic view of the citric acid production is given in Figure 5 (Benefit/cost ratio) and Table 6 (Cost/benefit ratio).

In this study, a considerable amount of citric acid was produced using red seaweed (*Gelidiella acerosa*) as the substrate by submerged fermentation. It contained 60% of carbohydrates. When the seaweed was added with the production medium, the pH slowly reduced at regular intervals from the initial pH of 6.5 to 3.5 on the tenth day. This change in pH indicated the production of acid in the production medium. Prior examinations report that the factors influencing citric acid production by fermentation incorporate the nutrient composition of the media, environmental conditions, lack of manganese (Rohr et al., 1983), types and concentration of sugars (Hossain et al., 1984), chelating effect on metal ions (Roukas and Kotzekiodou, 1997), ammonium nitrate, and aeration (Bayraktar and Mehmetoglu, 2000).

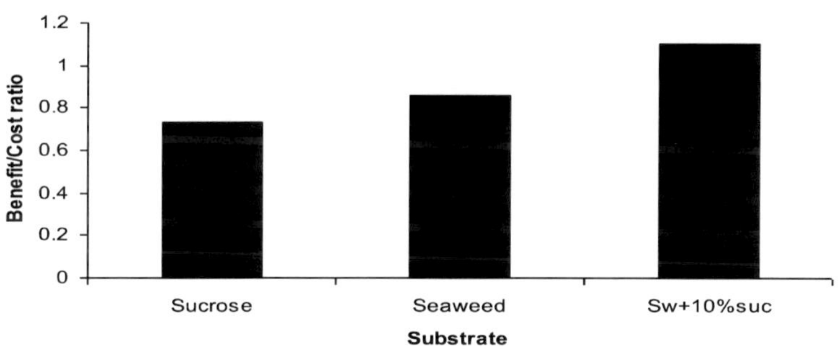

(Courtesy: Ramesh and Kalaiselvam, 2011)

Figure 5. Benefit/cost ratio of citric acid.

Table 6. The cost/benefit ratio of citric acid production in three different media

Source	Cost (C)		Benefit (B)		C/B ratio
	Input	Cost(Rs.)	Yield	Cost(Rs.)	
Sucrose (control medium)	1 kg	545	570 g	399	1.37
Seaweed	1 kg	200	250 g	175	1.14
Seaweed + 10% sucrose	1 kg	254	400 g	280	0.90

(Courtesy: Ramesh and Kalaiselvam, 2011)

The optimum time of incubation for maximal citric acid production differs with the organism and fermentation conditions utilized. In batch-wise fermentation of citric acid, the production begins after a lag phase of one day and arrived at greatest at the beginning of stationary phase or late-exponential phase. Further increment in the incubation period doesn't improve the citric acid production. It might be because of the age of fungus utilized and exhaustion of sugar contents in the culture broth (Arzumanov et al., 2000). As a matter of fact, the acid production begins with the underlying stage idiophase (somewhere in the range of 80 and 120 h) of fungal development. Right now, pH is commonly expressed at 5. The yield is slowly expanded, and it spans to the greatest rate at the late idiophase (180–220 h) (Guebel and Darias, 2001). Right now, pH must be beneath 3 so as to smother oxalic acid and gluconic acid formation (Crueger and Crueger, 2003). Right now, biomass of fungal culture has not been calculated in light of the fact that the fungal mycelium is changed into solid pellet forms in the control production medium; in other two media, the fungal mycelium is blended in with seaweed powder, and it's because of this the biomass isn't significant.

Temperature assumes a significant job in citric acid production. Temperature between 25–30°C is normally utilized for culturing of *A. niger*. As per Gerhardt et al. (1946), 30°C is the optimum temperature for citric acid production; however, temperature of medium increments above 30°C and therefore the biosynthesis of citric acid is diminished. It might be high temperature that can cause denaturation of enzyme-citrate synthase and amassing of other by-product acids, such as oxalic acid and enzyme catabolite repression (Panda et al., 1984), and it additionally restrains the culture advancement.

Results show that the concentration and kind of sugar impact the yield of citric acid production by *A. niger*, as noticed by Peksel and Kubicek (2003). In control production medium, the underlying pH 6.5 is decreased step by step to 1.5 during fermentation. Currently, sucrose is the substrate for citric acid production. The fungus *A. niger* uses the sugar compound and delivers 80g l^{-1} (57% citric acid), though seaweed alone in the production medium yields just 30g l^{-1} (22% citric acid). Sucrose generally has a low molecular weight, and it is promptly shipped into microbial cells for hydrolysis by intracellular enzymes (El-Holi and Al-Delaimy, 2003). This is the primary explanation behind utilizing sucrose-based substrate in citric acid production. Hossain et al. (1985) have additionally called attention to the fact that the chemical nature of sugar source markedly affects citric acid production by *A. niger*. Glucose, fructose, and maltose have likewise been utilized as substrates for citric acid production (Xu et al., 1989).

Begum et al. (1990) have discovered that the blend of two sugars in the medium at half of each improves the yields of citric acid. Angumeenal and Venkappayya (2012) reported that mixed culture fermentation yields a huge quantity of citric acid. When compared to a single culture medium, they obtained a heavy yield when the glucose-containing medium was supplemented with cadmium, chromium, lead, and molybdenum and inoculated with *A. niger* and *C. guilliermondii*. In the subsequent test, when the crude powder of seaweed is included at 10% of sucrose, 50 g l^{-1} (36% citric acid) is delivered as apparent by the pH decrease from 6.5 to 3.0. This might be because the carbohydrate, which is available in the seaweed joined with the sucrose, prompts the creation of citric acid. Shu and Johnson (1947) have indicated that a maximal citric acid creation rate is acquired at a range from 14 to 22% of the sugar in the medium; in some cases, the mix of two sugars (half of each) with the mutant strains improves the yields of citric acid.

Nitrogen is a limiting factor in the citric acid production. Nitrogen is generally provided as ammonium nitrate, which is metabolized during fermentation periods. Citric acid begins to show up when the nitrogen focus falls underneath a low limiting value. Apparently, the citric acid is created through carbon-storing cells.

Trace elements are the one of the most significant factors influencing the yield of citric acid fermentation. Specifically, the degrees of manganese, iron, copper, and zinc are very basic. On the off chance that the degrees of these trace elements are at an ideal level, different factors have less articulated impacts. On the other hand, the medium won't permit high production except if the trace component content is controlled cautiously. Shu and Johnson (1947) announced that the nearness of Mn and malt extract in the sporulation medium diminished the citric acid production in submerged fermentation, yet it did not do so in the surface culture fermentation. Noguchi and Johnson (1961) acquired the citric acid within the sight of constrained measures of iron and zinc; they additionally created 80% of citric acid from the sugars purged with chelating resins enhanced with zinc and iron. Wold et al. (1976) directed the citric acid collection in *A. niger* by the expansion of zinc and adenosine 3', 5'–cyclic monophosphate. Paul et al. (2019) recommended that high concentration of nutrient substrates, low nitrogen, and few quantities of trace elements for maintaining the increase dissolved oxygen and low pH are the best suitable conditions for high yield of citric acid by *A. niger*.

It is found from the examination that using seaweed alone as a sugar substrate for citric acid production by aging is lower (30 g l^{-1}) when contrasted with seaweed and 10% sucrose (50 g l^{-1}). The most extreme sum (80 g l^{-1}) of citric acid is delivered from sucrose. While making the seaweed powder using a mixer grinder, a tremendous measure of warmth is freed. This may meddle with the yield of citric acid when it is utilized as a substrate. It is proposed that keeping up a particular concentration and kind of sugars most likely is helpful to build the citric acid production. Further, a different or consolidated impact of basic environmental parameters and the mutant strain of *A. niger* for citric acid production should be considered cautiously.

Financially, the expense of chemically delivered and commercially accessible citric acid is 700 rupees per kilogram. (Sd-fine Chemicals Pvt. Ltd.). The principle substrate sucrose is regularly high cost at 545 rupees per kilogram. It yields the total production of 570g of citric acid. Simultaneously, one kilogram of seaweed (priced at 200 rupees) yields around 250g. Be that as it may, the media containing seaweed and 10% sucrose gives 400g at the expense of 254 rupees. This shows more advantage with ease when contrasted with the chemically synthesized citric acid. Along these lines, seaweed supplementation for citric acid production is an economically reasonable exercise.

Conclusion

There is a shortage for citric acid around the world. Microorganisms, especially *Aspergillus niger*, are a blessing as they satisfy the demand by creating citric acid through effectively utilizing accessible and cheap substrates. In the future, atomic-based strain improvement and upgrade systems in the microbial metabolism are important to accomplish more noteworthy statures in their creation by chemical industries.

References

Abd El Aty, A. A., Wehaidy, H.R. & Mostafa, F.A. (2014). Optimization of inulinase production from low cost substrates using Plackett–Burman and Taguchi methods. *Carbohydr Polym*, *102*, 261–268.

Almousa, A.A., Mohamed, N., Abd El-Ghany & Ashour, E.H. (2018). Citric Acid Fermentation by *Aspergillus niger. Journal of Innovations in Pharmaceutical and Biological Sciences (JIPBS)*, *5*(4), 20-37.

Angumeenal, A. R. & Venkappayya, D. (2012). An overview of citric acid production. *LWT - Food Science and Technology*. doi:10.1016/j.lwt. 2012.05.016.

Arzumanov, T.E., Shishkanova, N.V. & Finogenova, T.V. (2000). Biosynthesis of citric acid by *Yarrowia lipolytica* repeat-batch culture on ethanol. *Applied Microbiology and Biotechnology*, *53*, 525-529.

Bayraktar, E. & Mehmetoglu, U. (2000). Production of citric acid using immobilized conidia of *Aspergillus niger*. *Applied Biochemistry and Biotechnology*, *87*, 117-125.

Begum, A.A., Choudhury, N. & Islam, M.A. (1990). Citric acid fermentation by gamma ray induced mutants of *Aspergillus niger* in different carbohydrate media. *Journal of Fermentation Engineering*, *70* (4), 286.

Berovic, M. &Legisa, M. (2007). Citric acid production. *Biotechnology Annuual Review*, *13*, 303-43.

Bhargav, S., Panda, B.P., Ali, M. & Javed, M. (2008). *Solid-state Fermentation: An Overview*, *Chem. Biochem. Eng. Q.*, *22* (1), 49–70.

Couto, S.R. & Sanroman, M.A. (2006). Application of solid-state fermentation to food industry – a review *J. Food Eng.*, *76*, 291-302.

Crueger, W. & Crueger, A. (2003). Organic acids, In *Biotechnology: A textbook of Industrial Microbiology*, Second edition, Panima Publishing Corporation, New Delhi.

Dubey, R.C. (2001). *A Text Book of Biotechnology*, S. Chand & Company Ltd., New Delhi.

Elholi, M.A. & Al-Delaimy, K.S. (2003). Citric acid production from whey with sugars and additives by *Aspergillus niger. African Journal of Biotechnology*, *2* (10), 356-359.

Gerhardt, P., Dorrell, W.W. & Baldwin, I.L. (1946). Citric Acid Fermentation of Beet Molasses. *Journal of Bacteriology*, *52*, 555.

Grapulin, M.G. & Legisa, M. (1997). A spontaneous change in the intracellular cyclic AMP level in *Aspergillus niger* is influenced by the sucrose concentration in the medium and by light. *Applied and Environmental Microbiology*, *63* (7), 2844-2849.

Guc, S. & Erkmen, O. (2017). Citric Acid Production from Nontreated Beet Molasses by a Novel *Aspergillus niger* Strain: Effects of pH, Sugar and Ingredients. *J Food Microbiol Saf Hyg*, *2*, 122.

Guebel, D.V. & Darias, N.V.T. (2001). IFT Annual meeting – New Orleans, Louisiana.

Gupta, S. & Sharma, B.C. (1994). Continuous production of Citric acid from sugarcane molasses using a combination of submerged immobilized and surface stabilized cultures of *Aspergillus niger* KCU 520. *Biotechnology Letters*, *16* (6), 599 – 604.

Gupta, Treichel H., Shapaval, V., de Oliveira, L. A. & Tuohy, M.G. (2018). (Eds) *Microbial Functional Foods and Nutraceuticals*, First Edition. Published by John Wiley & Sons Ltd.

Haider, M. M. (2014). Citric acid production from carob pod extract by *Aspergillus niger. Journal of Pharmacy and Biological Sciences*, *9*, 112-116.

Hossain, M., Brooks, J. D. & Maddox, I. S. (1984). The effect of sugar source on citric acid production by *Aspergillus niger*. *Applied Microbiology and Biotechnology*, *19*, 393-397.

Hossain, M., Brooks, J. D. & Maddox, I. S. (1985). Galactose inhibition on citric acid production from glucose by *Aspergillus niger*. *Applied Microbiology and Biotechnology*, *22* (2), 98 - 102.

Kiel, H., Guvrin, R. & Henis, Y. (1981). Citric acid fermentation by *Aspergillus niger* on low sugar concentrations and cotton wastes. *Applied and Environmental Microbiology*, *42* (1), 1-4.

King, R. D. & Cheetham, P. S. J. (1987). Organic acids by fermentation: In *Food Biotechnology*. Elsevier Applied Science Publishers, Limited London, 273-308.

Kudzai, C. T., Ajay, K. & Ambika, P. (2016). Citric acid production by *Aspergillus niger* using different substrates. *Malaysian Journal of Microbiology*, *12*(3), 199–204.

Kunamneni, A., Permaul, K. & Singh, S. (2005). Amylase production in solid state fermentation by the thermophilic fungus *Thermomyces lanuginosus*. *Journal of Bioscience and Bioengineering*,*100*(2), 168-171.

Legisa, M. & Grapulin, M. G. (1995). Sudden substrate dilution induces a higher rate of citric acid production by *Aspergillus niger*. *Applied and Environmental Microbiology*, *61* (7), 2732-2737.

Max, B., Salgado, J. M., Rodríguez, N., Cortés, S., Converti, A.&Domínguez, J. M. (2010). Biotechnological production of citric acid. *Brazilian Journal of Microbiology*, *41*, 862–875.

Nitin, G. K., Deepali, M., Kiran, P., Priyanka, B. & Dhanke, P. (2017). A Review On Citric Acid Production And Its Applications. *International Journal of Current Advanced Research*, *6* (9), 5880-5883. DOI: http://dx.doi.org/10.24327/ijcar.2017.

Noguchi, Y. & Johnson, M. J. (1961). Citric acid fermentation of sugars purified with chelating resin. *Journal of Bacteriology*, *82*, 538-541.

Panda, T., Kundu, S. & Majumdar, S. K. (1984). Studies on citric acid production by *Aspergillus niger* using Indain molasses. *Journal of Microbiology*, *52* (2), 61-66.

Pandey, A. (2003). Solid state fermentation. *Biochemical Engineering Journal*, *13* (2-3), 81-84. https://doi.org/10.1016/S1369-703X(02) 00121-3.

Paul, P. E. V., Sangeetha, V. & Deepika, R. G. (2019). Emerging Trends in the Industrial Production of Chemical Products by Microorganisms. *Recent Developments in Applied Microbiology and Biochemistry*, Chapter 9 https://doi.org/10.1016/B978-0-12-816328-3.00009-X.

Peksel, A. & Kubicek, C. P. (2003). Effects of Sucrose concentration during citric acid accumulation by *Aspergillus niger*. *Turkish Journal of Chemistry*, *27*, 581-590.

Ramesh, T. & Kalaiselvam, M. (2011). An Experimental Study on Citric Acid Production by *Aspergillus niger* Using *Gelidiella acerosa* as a substrate. *Indian Journal of Microbiology*, *51*(3), 289–293. DOI 10.1007/s12088-011-0066-9.

Riviere, J. (1997). *Industrial applications of Microbiology*, Surrey University Press, London.

Rodrigues, C., Vandenberghe, L. P. S., Sturm, W., Dergint, D. E. A., de Carvalho, M. R. J. C. & Soccol, C. R. (2013). Effect of forced aeration on citric acid production by *Aspergillus* sp. mutants in SSF, *World Journal of Microbiology and Biotechnology*, DOI 10.1007/s11274-013-1397-y.

Rohr, M., Kubicek, C. P. & Kominek, I. (1983). Citric acid. In *Biotechnology*, *3*, VCH Publishers, Weinheim, Germany.

Roukas, T. & Kotzekiodou, P. (1997). Pretreatment of date syrup to increase citric acid production. *Enzyme Microbial Technology*, *21*, 273-276.

Sauer, M., Porro, D., Mattanovich, D. & Branduardi, P. (2008). Microbial production of organic acids: expanding the markets. *Trends in Biotechnology*, *26* (2), 100-108.

Shu, P. & Johnson, M. J. (1947). Effect of the composition of th esporulation medium on citric acid production by *Aspergillus niger* in submerged culture. *Journal of Bacteriology*, *54*, 161167.

Soccol, C. R., Vandenberghe, L. P. S., Rodrigues, C. & Pandey, A. (2006). New Perspectives for Citric Acid Production and Application. *Food Technology and Biotechnology*, *44*, 141–9.

Thomas, L., Larroche, C. & Pandey, A. (2013). Current developments in solid-state fermentation. *Biochemical Engineering Journal*, *81*, 146-161.

Tong, Z., Zheng, X., Tong, Y., et al. (2019). Systems metabolic engineering for citric acid production by *Aspergillus niger* in the post-genomic era. *Microbial Cell Factories*, *18*(1), 1–15.

Wold, W. S. M. & Suzuki, I. (1976). Regulation by zinc and adenosine 3,5, cyclic monophosphate on growth and citric acid accumulation I *Aspergillus niger*. *Canadian Journal of Microbiology*, *22*, 1093-1101.

Xu, D. B., Madrid, C. P., Rohr, M. & Kubicek, C. P. (1989). Influence of type and concentration of the carbon source on citric acid production by *Aspergillus niger*. *Applied Microbiology and Biotechnology*, *30*, 553-558.

Yigtoglu, M. (1992). Production of citric acid by fungi. *Journal of Islamic Academy of Sciences*, *5* (2), 100-106.

INDEX

A

B

C

D

E

F

G

H

I

K

L

M

T

U

W

X

Z